The Unified Neutral Theory of
Biodiversity and Biogeography

MONOGRAPHS IN POPULATION BIOLOGY

EDITED BY SIMON A. LEVIN AND HENRY S. HORN
Titles available in the series (by monograph number)

The Unified Neutral Theory of
Biodiversity and Biogeography

STEPHEN P. HUBBELL

PRINCETON UNIVERSITY PRESS

PRINCETON AND OXFORD
2001

Copyright ©2001 by Princeton University Press
Published by Princeton University Press,
41 William Street,
Princeton, New Jersey 08540
In the United Kingdom: Princeton University Press, 3 Market Place,
Woodstock, Oxfordshire OX20 1SY

Library of Congress Cataloging-in Publication Data

Hubbell, Stephen P., 1942–
    The unified neutral theory of biodiversity and biogeography /
Stephen P. Hubbell.
        p. cm.—(Monographs in population biology ; 32)
    Includes bibliographical references (p.   ).
    ISBN: 0-691-02129-5 (cloth : alk. paper)—ISBN 0-691-02128-7
(pbk. : alk.paper)   1. Biological diversity. 2. Biogeography. I. Title.
II. Series.

QH541.15.B56 H83 2001
578′.09—dc21   00-051637

This book has been composed in Baskerville

The paper used in this publication meets the minimum requirements
of ANSI/NISO Z39.48-1992 (R1997) (*Permanence of Paper*)

www.pup.princeton.edu

Printed in the United States of America

1   3   5   7   9   10   8   6   4   2

   3   5   7   9   10   8   6   4
(Pbk.)

In memory of
my father,
THEODORE HUNTINGTON HUBBELL
(1897–1989),
who taught me the love of nature,
and of
my wife,
LESLIE KILHAM JOHNSON
(1945–1997),
who taught me the nature of love

Contents

Preface

This is a book about a new theory of biodiversity in a geographical context. Work on this theory has been motivated by my conviction that better theories of biodiversity are urgently needed to inform our efforts to describe, manage, and protect it. Understanding biodiversity and its origin, maintenance, and loss on Earth is an issue of profound significance to the future of humanity and life as we know it.

In recent years, international attention to biodiversity issues has been growing. In my experience, however, too few people, including many of my distinguished academic colleagues and policymaker friends and acquaintances, fully grasp the enormity and urgency of this scientific and socioeconomic problem. In part because of this ignorance, investment in the science of biodiversity lags far behind investment in biomedical research. We know how to treat a cancer patient with monoclonal antibodies and genetically engineer pest resistance in crop plants, but we do not know how many species inhabit the Earth or even a small part of it—even to the nearest order of magnitude (May 1988). We know even less about how and where most species on Earth originate, live, and die. Many of my colleagues in other fields are surprised to learn that the study of biodiversity is still largely in a Linnaean phase of discovering and naming new species. Although our tools are more advanced, in many ways the science of biodiversity is not much farther along than medicine was in the Middle Ages. We are still at the stage, as it were, of cutting open bodies to find out what organs are inside.

The low investment in and slow pace of biodiversity research might be tolerable were it not for the overwhelmingly rapid destruction of the natural world. Without hyperbole we can truthfully say that we are almost out of time to save much of the diversity of life on Earth. We humans already sequester an astonishing 40% of the entire terrestrial primary production of the Earth for our own selfish use, and this percentage is increasing every year (Vitousek et al. 1986, Tilman 2000). Capturing such an enormous fraction of the Earth's natural productive capacity has come at a huge cost in terms of loss of natural habitat and reduction in the viability or outright extinction of species. In this book, I make a great deal of the argument that living nature is forever locked in a life-or-death, zero-sum game for limiting resources. All the rest of life on Earth is being squeezed into an ever-shrinking fraction of the total planetary resource pool. This compression of all nonhuman life into ever smaller bits of habitable land and sea is not sustainable and is highly destabilizing. Consider tropical moist forests, which now cover less than 7% of the global land surface area but are thought to protect as many as three quarters of all surviving species on Earth (Wilson and Peter 1988). About 20% of the world's original tropical moist forest has already been completely destroyed, and another 40% has been seriously degraded, most of it in the last century. The remainder is being cut down at a rate estimated at 1% to 2% per year (Myers 1988).

In view of the genuine possibility of a global collapse of biodiversity in the near future, it is unconscionable that we still have no serviceable general theory of biodiversity. The development of such a theory should be made a national and international research priority. Some of my colleagues are pessimistic. They seem to think that the search for such a theory is a quixotic waste of time, that there are no general principles to be elucidated, and that the study of biodiversity is little more than a loose collection of unique evolutionary events and natural history stories. I believe that they are wrong, and fortunately not everyone, including me, is so pessimistic (e.g., Brown 1995, Avise 1999, Ritchie and Olff 1999).

One reason why this pessimism exists is that the term *biodiversity* has been coopted by the policy community, where the term has become too inclusive. In policy discussions, biodiversity covers an enormous and heterogeneous array of subjects, scales, and questions. "Biodiversity" in policy parlance is the sum total of all biological variation from the gene level to single-species populations of microbes to elephants, and multispecies communities and ecosystems to landscape and global levels of biotic organization. In some usages, it also includes all ecological interactions within and among scales of biological organization. In this book I seek a general theory of biodiversity, but one that is more narrowly defined and in the context of the classical scientific discipline of ecology.

This book is the outgrowth of a graduate seminar and a series of undergraduate lectures on biodiversity and biogeography at Princeton University. However, the germ of the ideas presented here can be traced to my thirty-year-old fascination with the origin and maintenance of high tree species diversity in tropical forests, an interest that launched several large-scale studies of tropical forest diversity and dynamics in the late 1970s. The earliest version of the present theory was published over twenty years ago in a paper in *Science* on tree diversity in a tropical dry forest in Costa Rica (Hubbell 1979).

At some point now lost to memory, while teaching the theory of island biogeography, I wondered what would happen if a process of speciation were incorporated into the theory. I did not actually attempt this in the formal context of a mathematical theory until 1995. However, when I eventually did the math, I was completely unprepared for what happened next. Adding speciation unexpectedly resulted in a unification of the theories of island biogeography and relative species abundance—theories that heretofore have had almost completely separate intellectual histories. Exactly what I mean by unification will become clear later in the book. But among other things, this unified theory generates a truly remarkable dimensionless number. In the theory, this number controls not only species richness on islands and the mainland, but also the island and mainland distributions of relative species abundance, species-area relationships, and even phylogeny. Because of its ubiquity and centrality in the theory, I have christened it the *fundamental biodiversity number*. Equally exciting from a practical point of view, this biodiversity number can be estimated from relative species abundance data, and from this we can potentially obtain information about speciation rates and the sizes of source metapopulations and metacommunities. For the first time, we have a formal theory that connects speciation to biogeography and large-scale patterns of biodiversity.

I dedicate this book to the memory of my father, Dr. Theodore Huntington Hubbell. My father, director of the Museum of Zoology at the University of Michigan, entomologist and evolutionary biologist *extraordinaire*, first introduced me to "biodiversity" before it was called such, on many field trips in the United States, Mexico, and Central America. He is a hard act to follow.

I also dedicate this book to the memory of my wife of twenty-one years, Dr. Leslie Kilham Johnson, as scientist, scholar, artist, teacher, mother, and lover. Midway through the writing of this book, Leslie died after a two-year struggle with breast cancer. In the last six months of her life, my colleagues in the Ecology and Evolutionary Biology Department at Princeton kindly took over all of Leslie's and my departmental and teaching responsibilities so that I could be home with her full time. I am immensely grateful to them for this generosity and compassion. During this time, Leslie was incredibly brave and she urged me to finish the book.

I am now in love with and married to Dr. Patricia Adair Gowaty, who has my immense gratitude for saving my life and providing the love and encouragement I needed to complete this book, five years after I began. She has also read and reread draft after draft, and the book is greatly improved because of her consummate editorial skills and her intellectual capacity to cut right to the essentials.

I am particularly grateful to four people who read the entire book and gave me constructive suggestions for its improvement: Jorge Ahumada, Jared Diamond, Henry Horn, and Mark McPeek. I am also grateful to the many others who have read sections of the manuscript along the way or have otherwise provided feedback, data, or simply old fashioned encouragement. In alphabetical order, they include Peter Ashton, John Bonner, Walt Carson, Rick Charnov, Peter Chesson, Liza Comita, Rick Condit, Laurence Cook, Andy Dobson, John Endler, Warren Ewens, Robin Foster, Kyle Harms, Allen Herre, Jeremy Jackson, Jeff Klahn, Nancy Knowlton, Russ Lande, Bert Leigh, Simon Levin, Karen Masters, Helena Muller-Landau, Sean O'Brien, Steve Pacala, John Pandolfi, David Peart, Ron Pulliam, Bob Ricklefs, Mark Ritchie, Ira Rubinoff, Kalle Ruokolinen, Henry Stevens, R. Sukumar, I-Fang Sun, Vince Tepedino, John Terborgh, Cam Webb, Kirk Weinmiller, Peter White and Joe Wright.

I also thank the National Science Foundation, the John D. and Catherine T. MacArthur Foundation, the Andrew K. Mellon Foundation, the Pew Charitable Trusts, the John Simon Guggenheim Foundation, and numerous other foundations, organizations, and individuals for supporting or assisting in the research on tropical forests that has provided much of the data discussed in this book. I also give special thanks to Ira Rubinoff, director of the Smithsonian Tropical Research Institute, and Elizabeth Losos, director

The Unified Neutral Theory of
Biodiversity and Biogeography

MacArthur and Wilson's Radical Theory

This is a book about a new general theory of biodiversity in a geographical context. I define *biodiversity* to be synonymous with species richness and relative species abundance in space and time. *Species richness* is simply the total number of species in a defined space at a given time, and *relative species abundance* refers to their commonness or rarity. This is a less inclusive definition of biodiversity than is commonly used in policy circles, but more in keeping with the classical discipline of ecology as the scientific study of the distribution and abundance of species and their causes. Fragments of a general theory of biodiversity abound in ecological theories of island biogeography, metapopulations, and relative species abundance; but in my opinion, there have not yet been any really successful syntheses.

Among the kinds of diversity patterns I seek to explain with this new theory are those illustrated in figure 1.1. This graph shows patterns of relative species abundance in a diverse array of ecological communities, ranging from an open-ocean planktonic copepod community, to a tropical bat community, to a community of rainforest trees, to the relative abundances of British breeding birds. Each line is a plot of the logarithm of the percentage relative abundance of species on the *y*-axis against the rank in abundance of the species on the *x*-axis, from commonest at the left (low ranks) to rarest on the right (high ranks). The curves differ in many ways, including species richness, the degree of

3

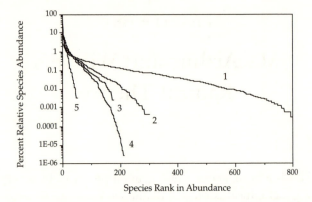

F<small>IG</small>. 1.1. Patterns of relative species abundance in a diverse array of ecological communities. Species in each community are ranked in percentage relative abundance from commonest (*left*) to rarest (*right*). The percentage relative abundance is log transformed on the *y*-axis. 1: Tropical wet forest in Amazonia. 2: Tropical dry decidu-ous forest in Costa Rica. 3: Marine planktonic copepod community from the North Pacific gyre. 4: Terrestrical breeding birds of Britian. 5: Tropical bat community from Panama.

dominance of the community by common species, and the number of rare species each community contains. Neverthe-less, the relative abundance distributions of this heteroge-neous collection of communities all have a curiously similar shape. Some are steeper, and some are shallower, but all of the distributions basically exhibit an S-shaped form, bend-ing up at the left end and down at the right end. Is there a general theoretical explanation for all of these curves? Can we hope even to find a quantitative theory that accurately predicts the relative abundances of the individual species in each of these distributions? I believe so, and I also believe a general theory exists for much more. The development of such a theory is the central theme of this book.

Before proceeding, it is important to issue some caveats and define some terms. Although this purports to be a general theory of biodiversity, in fact it is a theory of within-trophic-level diversity. There are many aspects of community organization that currently lie beyond the theory's scope, such as the trophic organization of communities (e.g., Hairston et al. 1960, Oksanen 1988), or what controls the number of trophic levels (e.g., Cohen 1978, Pimm 1982, 1991), or how biodiversity at one trophic level affects diversity on other trophic levels (e.g., Paine 1966, Janzen 1970, Connell 1971, Pimm 1991, Holt 1977, Strong et al. 1984).

For present purposes, I define an *ecological community* as a group of trophically similar, sympatric species that actually or potentially compete in a local area for the same or similar resources. Examples might be tree species in a forest, or zooplankton grazing on phytoplankton in a lake. This definition might appear to be closer to what ecologists commonly refer to as an *ecological guild*. However, as I explain in chapter 10, I believe that the theory will often apply more generally to species that are in a particular trophic level but would otherwise be classified in different ecological guilds. While not complete, a theory of biodiversity within trophic levels would nevertheless be a major advance because most biodiversity resides within rather than between trophic levels (i.e., there are many more species than trophic levels). Also, grouping species into trophically similar classes is perhaps the most logical, natural, and tractable way to address questions of species diversity. Not surprisingly, this is the domain for most of the theory about niche partitioning in community assembly (Tokeshi 1993, 1997, 1999). I will also use the term *metacommunity* when applying the theory to biodiversity questions on large, biogeographic spatial scales and on evolutionary timescales. The *metacommunity* consists of all trophically similar individuals and species in a regional collection of local communities. However, unlike species in the local community, metacommunity species may not actually compete because of separation in space or time.

The theory presented here is constructed on the foundation of the equilibrium theory of island biogeography and owes a great debt to the original insights of MacArthur and Wilson (1963, 1967). It is also an unabashedly neutral theory. I examine the theoretical consequences of assuming that ecological communities are structured entirely by ecological drift, random migration, and random speciation. By *neutral* I mean that the theory treats organisms in the community as essentially identical in their per capita probabilities of giving birth, dying, migrating, and speciating. This neutrality is defined at the *individual* level, not the species level, a distinction whose importance will be explained shortly. The term *ecological drift* is not currently in widespread usage, but it is essentially identical to a concept already familiar to most ecologists as *demographic stochasticity*. While the assumption of complete neutrality is patently false, few ecologists would deny that real populations and communities are subject not only to physical factors and biotic interactions, but also to demographic stochasticity. To study ecological drift theoretically, it is easier to make the assumption of per capita ecological equivalence—at least to begin with. In the plant literature, the notion of ecological equivalence is by no means a new idea (e.g., Hubbell 1979, Goldberg and Werner 1983, Shmida and Ellner 1984).

Before proceeding, I need to be more precise about the meaning of *neutrality* as used in this book. Despite its moniker, the concept of neutrality actually has many meanings in the literature. To most people, the word *neutral* congers up the qualitative notion of "nothing going on." But exactly what people mean by this phrase often turns out to differ from one person to the next. I use *neutral* to describe the assumption of per capita ecological equivalence of all individuals of all species in a trophically defined community. This is a very unrestrictive and permissive definition of *neutrality* because it does not preclude interesting biology from happening or complex ecological interactions from taking place among individuals. All that is required is that all individuals of every species obey exactly the same rules of ecological engagement. So, for example, if all individuals and species enjoy a frequency-dependent advantage in per capita birth rate when rare, and if this per capita advantage is exactly the same for each and every individual of a species of equivalent abundance (e.g., Chesson and Warner 1981), then such a theory would qualify as a *bona fide* neutral theory by the present definition. The theory of island biogeography is a neutral theory, but it nonetheless has "ecological rules" that govern the rates of immigration and extinction of assumed identical species to and from islands. Thus, *the essential defining characteristic of a neutral theory in ecology is not the simplicity of its ecological interaction rules, but rather the complete identity of the ecological interaction rules affecting all organisms on a per capita basis.* In the present book, I will consider only one class of all possible neutral theories. I examine the consequences of assuming that population and community change arises only through ecological drift, stochastic but limited dispersal, and random speciation. Furthermore, I will consider only one possible mechanism of ecological drift, namely one that I have named zero-sum ecological drift, which I will define and discuss in chapter 3.

Theory aside, it may be hard for many ecologists to accept that ecological drift might actually be important in natural populations and communities. The physicist Heinz Pagels (1982) once observed that there seem to be two kinds of people in the world. There are those who seek and find deterministic order and meaning, if not purpose, in every event. And then there are those who believe events to be influenced, if not dominated, by intrinsically inscrutable, and meaningless, random chance. One of the intellectual triumphs of twentieth-century physics was to prove that both views of physical nature are simultaneously true and correct, but on very different spatial and temporal scales. This dualism remains difficult for many people to accept, however. Even Albert Einstein, whose own work overthrew classical Newtonian determinism, never fully accepted

7

quantum mechanics and Heisenberg's uncertainty principle (Kevles 1971). Physicists have had to be content with the order discovered in the laws of quantum probability, not in the indeterminism of the quantum events themselves. Ironically, the discovery of deterministic chaos in natural systems has undermined even the once rock-solid faith that determinism ensures predictability (Gleick 1987).

Somewhat analogous philosophical dualisms also run through population genetics and ecology. In population genetics, a long-standing debate has persisted over whether most change in gene frequencies results from random, neutral evolution or from natural selection (Crow and Kimura 1970, Lewontin 1974). In ecology, there are two conflicting world views on the nature of ecological communities which were brought into stark relief by MacArthur and Wilson's theory, although perhaps only in hindsight. The mainstream perspective is what I will call, at the risk of caricature, the *niche-assembly perspective.* This view holds that communities are groups of interacting species whose presence or absence and even their relative abundance can be deduced from "assembly rules" that are based on the ecological niches or functional roles of each species (e.g., MacArthur 1970, Levin 1970, Diamond 1975, Weiher and Keddy 1999). According to this view, species coexist in interactive equilibrium with the other species in the community. The stability of the community and its resistance to perturbation derive from the adaptive equilibrium of member species, each of which has evolved to be the best competitor in its own ecological niche (Pontin 1982). Niche-assembled communities are limited-membership assemblages in which interspecific competition for limited resources and other biotic interactions determine which species are present or absent from the community.

The other world view might be dubbed the *dispersal-assembly perspective.* It asserts that communities are open, nonequilibrium assemblages of species largely thrown together by chance, history, and random dispersal. Species come and go, their presence or absence is dictated by random dispersal and stochastic local extinction. The theory of island biogeography is an example of such a theory. It asserts that the species in island communities are put there solely by dispersal, i.e., island communities are dispersal assembled, not niche assembled (Hubbell 1997). Although dispersal-assembly theories do not have to be neutral, most are, including MacArthur and Wilson's theory. It is neutral because their famous graphical model assumes that all species are equal in their probabilities of immigrating onto the island, or of going extinct once there. The neutrality of the theory of island biogeography is not always appreciated, however, because MacArthur and Wilson were themselves not completely explicit on this point—about which I will have more to say in a moment.

Neutral theories have had a checkered history in community ecology of creating more heat than light (Strong et al. 1984). Many of the early attempts at constructing neutral models were direct outgrowths of the theory of island biogeography, and some had statistical and other problems that tended to bring misdirected discredit down on the theory of island biogeography itself. The absence of a good neutral theory in community ecology is due in part to the fact that what little neutral theory exists has not been very convincing (Caswell 1976, but see Gotelli and Graves 1996). However, the lack of neutral theory in ecology is also due in no small measure to widespread and, in my opinion, counterproductive resistance to such theories among ecologists.

A recent example of resistance to neutral ideas can be found in an otherwise excellent review by Chesson and Huntly (1997), who argue against the importance of ecological drift (although they did not call it by that name) in structuring ecological communities. The nub of their argument is captured in the following quote: "The unfortunate failure to emphasize that, at some spatial or temporal scale, niche differences are essential to species coexistence has allowed logical inconsistencies in the ideas . . . to remain unnoticed" (p. 520). As I will repeatedly demonstrate in this book, however, niche differences are not essential to coexistence, if by "coexistence" we mean the persistence in sympatry of species for geologically significant lengths of time. There is no logical inconsistency in the argument presented here. I prove in chapter 5 that a biodiversity equilibrium will arise between speciation and extinction that results in a long-term, steady-state distribution of relative species abundance. While it is true that all these species are transient, transit times to extinction will be measured in millions to tens of millions of years for most species that achieve even a modest level of total global abundance (chapter 8). Although there is undeniable evidence for niche differentiation among real species in many trophic guilds, this differentiation is not at all essential for coexistence on the timescales usually discussed by ecologists. As I will show, distinguishing the predictions of ecological drift from those made by niche-based theories that predict indefinite species coexistence will often be empirically difficult.

So I believe that we should seriously question *why* our theories in ecology always start with the presumption of the indefinite coexistence of species. All the evidence of which I am aware supports entirely the opposite conclusion—namely, that all species are transient and ultimately go extinct. It seems to me to be a big problem to erect the mathematical expectation of indefinite species coexistence as a sine qua non criterion for whether an ecological theory is acceptable or not. In this entire discussion we seem to have lost sight of the fact that ecology lacks a good *operational* definition of coexistence. The axiomatic premise of coexistence is the real reason why there is virtually no connection between theoretical and empirical discussions of the coexistence question. The loss of a species from a local community is a relatively commonplace observation. However, the number of cases in which local extinction can be definitively attributed to competitive exclusion is vanishingly small. We no longer need better theories of species coexistence; we need better theories for species presence-absence, relative abundance and persistence times in communities that can be confronted with real data. In short, it is long past time for us to get over our myopic preoccupation with coexistence.

My goals in this book are threefold: first, to develop a formal theory for ecological drift to see "how far we can get with it"; second, to recognize in ecology—as population genetics did quite some time ago—that ecological drift as well as biotic interactions (read genetic drift and selection for population genetics) are *both* potentially important to the assembly and dynamics of ecological communities; and third, to attempt to dispel the pervasive resistance of ecologists to neutral theory by demonstrating its considerable predictive power. Indeed, an intriguing feature of this theory is how surprisingly well it works and how unexpectedly rich it is in nonobvious testable predictions. I will endeavor to show that a neutral theory is not only possible for the origin and maintenance of biodiversity, but also that it generates a quantitative theory for relative species abundance, species-area relationships, and the landscape-level distribution of biodiversity. It also makes testable predictions about modes of speciation and patterns of phylogeny and phylogeography. The theory predicts that different modes of speciation will leave different biodiversity and biogeographic signatures in the distribution of metacommunity relative species abundance (chapter 8).

Neutral theories exist for many ecological patterns and processes at various spatio-temporal scales (Gotelli and Graves 1996). However, the neutral theory presented here is unique to my knowledge in unifying many of them into a single theory. It is the incorporation of speciation into the theory of island biogeography that enables the unified theory to predict relative species abundance from local to

11

global biogeographic scales. This unified theory also generates a dimensionless biodiversity number, θ, which appears to be fundamental in the sense that it crops up everywhere in the theory at all spatio-temporal scales (chapters 5, 6, 8). Quite remarkably, at least for now, the unified neutral theory does a better job of explaining patterns of biodiversity, relative species abundance, species-area relationships, and phylogeny than current niche-assembly theory does. This state of affairs is likely to change as more synthetic theories develop that include ecological drift as well as niche differences among species. However, regardless of the form that future ecological theories ultimately take, it will no longer be acceptable for these theories to ignore ecological drift.

I begin with a brief review and critique of the original theory of island biogeography, which is the intellectual cornerstone of the unified neutral theory. MacArthur and Wilson erected their theory in part to explain the puzzling observation that islands nearly always have fewer species than do sample areas on continents of the same size. Why is this so? They reasoned that perhaps extinction rates on islands would be higher because of smaller average population sizes (small populations are more extinction prone). Then, once island populations went extinct, it would take the same species longer to recolonize the island than it would take them to disperse among adjacent areas on the mainland. Thus, other things being equal, species would spend a smaller fraction of total time resident on a given island than in the same-sized area of the mainland. Given these assumptions, i.e., a higher island extinction rate and a lower reimmigration rate, one then predicts a lower steady-state number of species on islands than in same-sized areas on the mainland. MacArthur and Wilson captured this simple equilibrium idea in a now famous graph familiar to nearly every high school biology student (fig. 1.2).

By most yardsticks, MacArthur and Wilson's (1963, 1967) simple and intuitive theory of island biogeography has been

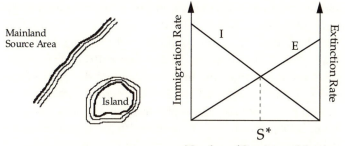

FIG. 1.2. MacArthur and Wilson's familiar equilibrium hypothesis for explaining the number of species on islands as a dynamic equilibrium (S^*) between the rate of immigration of new species onto the island and the rate of extinction of species already resident on the island. I = immigration rate; E = extinction rate.

a phenomenal and resounding success. The theory revived and sustained a broad interest in questions of biodiversity and biogeography (MacArthur 1972). It spawned a major growth industry in conservation biology by its application to issues such as reserve design (Diamond and May 1975) and the estimation of extinction rates (e.g., Levins 1978, Diamond 1972, 1984, Terborgh 1974, Terborgh and Winter 1980, May and Lawton 1995). It also inspired the fundamental paradigm for the emerging discipline of metapopulation biology (Levins 1969). The literature on island biogeography and allied subjects fills the ecological journals, and in my estimation now numbers in the neighborhood of a thousand papers.

By other measures, however, the theory has been a disappointment, not so much through any fault of its own, as some have charged (e.g., Slobodkin 1996, Hanski and Simberloff 1997), but largely because subsequent development of the theory has languished. Close to four decades have elapsed since MacArthur and Wilson (1963) published their seminal paper in *Evolution*, yet there have been only sporadic theoretical advances on the community-level questions that were addressed by the original theory and that have built directly on the foundation of the original insight (e.g., May 1975, Brown and Kodric-Brown 1977). Even more time has passed since the first, independent discovery of the theory by a Cornell graduate student named Monroe, who was studying butterfly biogeography in the Antilles (Monroe 1948, 1953, Brown and Lomolino 1989). In recent years there has been an explosion of empirical and theoretical research on metapopulation biology (Hanski and Gilpin 1997), but this opus consists almost entirely of work on the population dynamics of single species over a set of discrete habitat patches, not on the spatio-temporal dynamics of whole ecological communities.

There is no mystery why so little theoretical development of the equilibrium theory of island biogeography has occurred. MacArthur and Wilson's theory was—and, to a large extent, remains—a radical departure from mainstream thinking in contemporary community ecology. In its fundamental assumptions it is a neutral theory that asserts that island communities are dispersal assembled, not niche assembled. It is something of a misnomer to describe island biogeography as an equilibrium theory: it can only be narrowly construed as such. In this narrow sense, it predicts a steady-state number of species on islands under a persistent rain of immigrant species from mainland source areas (fig. 1.2). However, in contrast to niche-assembly community theory, it does not predict a stable assemblage of particular taxa. It predicts only a diversity equilibrium, not a taxonomic equilibrium.

The theory of island biogeography was radical because it broke away from the conventional neo-Darwinian view of ecological communities as coadapted assemblages of niche-differentiated species residing at or near adaptive and demographic equilibrium. In its place it erected a brave new world view in which ecological communities are seen as

14

in turmoil, in perpetual taxonomic nonequilibrium, under-going continual endogenous change and species turnover through repeated immigrations and local extinctions. These turnovers need not be especially rapid, however, and species can coexist for long periods in slowly drifting mixtures and in shifting relative abundances. The theory was all the more remarkable because it was elaborated by MacArthur himself, the leading ecological theorist of his day and champion of the dominant, niche-based equilibrium view.

If communities are largely accidental collections of species whose biogeographic ranges happen to overlap for histori-cal and individualistic reasons, then it follows that species in communities are not highly coadapted or codependent. Setting aside obligate mutualisms and host-parasite relation-ships, which are almost nonexistent between species within the same trophic level (the domain of the present the-ory), then species are rarely so dependent on one another that they cannot persist in the community without particu-lar other species. This view does not deny the obvious exis-tence of niche differentiation. However, it ascribes much less importance to niche in regulating the relative abun-dance and diversity of species in the community. Niche dif-ferentiation, according to this view, is seldom the result of pairwise competition from species sharing similar resource requirements—which helps to explain the apparent rarity of character displacement in nature (Brown and Wilson 1956, Grant 1972, 1975, 1986). Rather, niche differentiation reflects the time-averaged history of the ever-changing biotic and abiotic selective environments to which the species ancestral lineages were exposed during their long, individu-alistic geographic wanderings, the ghost of competition past (Connell 1980).

MacArthur and Wilson's theory raised the possibility that history and chance alone could play an equal if not larger role in structuring ecological communities than do niche-based assembly rules. The idea that random dispersal

15

and ecological drift could be important in structuring communities was by no means new. Dispersal-assembly theories date back at least to the seminal book, *Age and Area, A Study of Geographic Distribution and Origin in Species*, by a Cambridge University professor J. C. Willis (1922). Similar ideas were independently developed by Gleason (1922, 1926, 1939). However, the success of MacArthur and Wilson's theory brought these old ideas renewed credibility and attention.

According to island biogeography theory, it does not matter which species contribute to balancing immigration and extinction rates on any given island. The only state variable in the model is the number of species on the island. All species in the original theory are treated as identical. Without this assumption, the model's reduction of island community dynamics to counting species does not logically work. Various embellishments on the basic theory do not change this fact. For example, downwardly concave immigration and extinction curves were added to create a more "interactive" version of the theory (Simberloff 1969; fig. 1.3). This makes late-arriving species experience lower successful immigration rates and higher extinction rates. However, this modification does not alter the basic fact that any species arriving late, regardless of whether it is a good colonizer or competitor, will exhibit the same rate changes. Likewise, all species respond in an identical manner to varying the size of the island and its distance from the mainland source area. Other modifications include the "rescue effect" proposed by Brown and Kodric-Brown (1977), who noted that immigration will often interact with and reduce local extinction rates, particularly in local communities in continuous habitats; but again this addition does not alter the fundamental assumptions of the theory.

MacArthur and Wilson's dispersal-assembly hypothesis, when stripped to its bare essentials of total neutrality and species substitutability, will undoubtedly seem extreme to most people. Indeed, it is not clear whether MacArthur

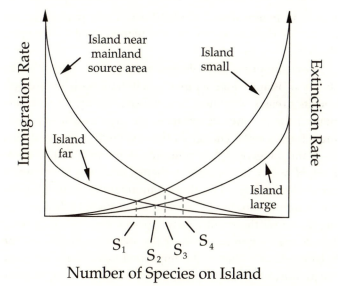

FIG. 1.3. Various enhancements to the basic equilibrium hypothesis of MacArthur and Wilson do not change the dispersal assembly assumption underlying the model. Downwardly bowing immigration and extincton curves were added to characterize the effects of competition on these rates, but all species, whether early or late colonizers, good or bad competitors, experience the same changes in rates. Similarly, the effects of island distance from the mainland and island size on immigration and extinction rates, respectively, operate equally on all species.

and Wilson fully appreciated the implications of this radical assumption. A majority of their 1967 monograph was devoted to discussing such topics as species differences in colonization strategies, causes of species differences in extinction rates, temporal patterning in the order in which species would successfully establish, and so on—all differences *forbidden* by their model! Although MacArthur and Wilson (1967) wrote about traditional ecological processes such as competition, the actual parameters of their model were immigration and extinction rates, distance from

17

mainland source areas and island size. These parameters are absent, or virtually so, from most niche-assembly ecological theories.

The assumption of ecologically equivalent species in the theory of island biogeography is simultaneously its strength and its greatest weakness as a neutral theory. The present neutral theory owes its existence in large measure to making a fundamental change in this assumption. In the theory of island biogeography, neutrality is defined at the *species* level. However, in the present theory, neutrality is defined at the *individual* level. This distinction might seem to be subtle and unimportant, but in fact it is key to successfully developing a neutral theory of relative species abundance. Making this change eliminates one of the major objections to the original theory. One of the important ways that species are *not* identical is in their relative species abundances. Relative species abundances, for example, strongly affect the average time it takes a species to go extinct through ecological drift. With this change, there is no longer any need to specify the extinction rate as a parameter—as was necessary in the theory of island biogeography—because the extinction rate can now be predicted by the theory as a function of population size. The ramifications of differences in relative species abundance are the focus of many of the theoretical explorations in this book.

A separate issue from the neutrality assumption is the fact that the theory of island biogeography is conceptually incomplete in a number of important regards. From a biogeographer's perspective, it is incomplete because it embodies no mechanism of speciation. Although species can appear and disappear from islands or habitats in the theory, this is a migration- and local extinction-driven phenomenon. No new species are allowed to originate on islands or in the source area. From a community ecologist's perspective, the theory is incomplete in large part because, as mentioned, it does not predict the relative abundance of species, only species richness. Relative abundance theory is briefly touched upon in MacArthur and Wilson's (1967) monograph with respect to the species-area relationship. However, the expected equilibrium distribution of relative species abundance on islands was not derived from the first principles of the theory.

The unified neutral theory of biodiversity and biogeography is a conceptual advance over either the theories of island biogeography or relative species abundance taken separately. In current theories of relative species abundance, the number of species in the community is a free parameter that cannot be derived from first principles (Motomura 1932, MacArthur 1957, 1960, Fisher et al. 1943, Preston 1948, 1962, Cohen 1978, Sugihara 1980) (chapter 2). In the unified theory, the equilibrium number of species is a prediction—as in the theory of island biogeography—but so is relative species abundance.

MacArthur and Wilson devoted a large portion of their 1967 monograph to discussing the relationship between island population size and risk of extinction. Without a theory of speciation and relative species abundance, however, they were unable to make headway on many other issues of central importance to community ecology and conservation biology, including expected abundances of species on islands and in the metacommunity and their variances, species incidence functions and times to extinction and recolonization, patterns of island and metacommunity dominance and diversity, species-individual and species-area relationships, and the spatial covariance of populations and of community composition. Much progress has been made on theory for some of these problems individually (e.g., May 1975, Caswell 1976, Coleman 1981, Quinn and Hastings 1987, Caswell and Cohen 1991, Hanski and Gilpin 1997, Durrett and Levin 1996). In the unified theory, many of these problems are now analytically tractable.

Before developing the unified neutral theory, it is useful to give a very brief intellectual history of the niche-assembly and dispersal-assembly perspectives. It is no accident that the perspective to which a person adheres can largely be predicted by the scale on which the person works. Most proponents of niche assembly come out of a strong neo-Darwinian tradition, which focuses on the lives of interacting individuals and their fitness consequences. The concept of niche follows naturally and logically as the population-level summation of the individual adaptations of organisms to their biotic and abiotic environments. Indeed, most ecologists after Grinnell (1917) explicitly grounded niche formalism in the language of fitness and intrinsic rates of increase (Whittaker and Levin 1975).

One consequence of a focus on adaptation and niche assembly has been a tendency to accept an equilibrium and a relatively static view of niches and ecological communities (see critiques by Weins 1984, Pimm 1991). Equilibrium thinking has been aided and abetted by mathematical ecology, which seeks to predict the ecological balance among niche-differentiated competing species, predators and their prey, and so on (Pontin 1982, Weiher and Keddy 1999). For reasons of tractability and hoped-for generality, most theory in population and community ecology focuses on equilibrium analyses and small perturbations around equilibria (Levins 1968, 1975, MacArthur 1972, May 1973, 1981, Maynard Smith 1968, 1974, Pielou 1977, Tilman 1982, Rose 1987). Similarly, theory in evolutionary ecology has elaborated extensively on the idea of the evolutionarily stable strategy (ESS), a concept that assumes the existence of non-invasible adaptive equilibria (Maynard Smith 1968, 1974, Krebs and Davies 1978, Charnov 1982).

This focus on individual variation in fitness, adaptation and niche has led naturally to small-scale, short-term experimental studies of processes of competition, selection and adaptation. Tilman (1989), for example, reviewed several hundred studies of plant competition and found that nearly half of all studies were on a spatial scale of a square meter or less, and three quarters were done in plots of less than 10×10 m. Only 15% of the studies lasted longer than 3 years, much less outlasted the research career of the original investigator.

Proponents of dispersal assembly, on the other hand, typically work on much larger spatial and temporal scales, using biogeographic or paleoecological frames of reference. Their approach is less experimental and more analytical of large-scale statistical patterns than the niche-based approach, but it is no less scientifically valid (Brown 1995). MacArthur once compared biogeography to astronomy, and quipped that no one has ever faulted astronomy for not being experimental (Rosenzweig 1995). A consequence of working at large spatial and temporal scales is a tendency to be impressed by how spatially variable and ephemeral ecological communities are. For example, the composition and species mixtures of plant communities change greatly from small scales (Whittaker 1956, 1967) to large scales (Gleason 1926, 1939), with few if any sharp community boundaries. On the basis of these patterns, Gleason promulgated his "individualistic" concept of plant distribution, arguing against the extreme niche-assembly view of his contemporary, Frederick Clements (1916). Clements believed that the community was literally a "superorganism," that species were its organs and succession its ontogeny. He argued that each species had an essential role to play in preparing the way for the next seral stage in the succession toward the equilibrium or "climax" plant community. Clements's superorganism theory was discredited by Tansley (1935) on the grounds that it was inconsistent with natural selection at the individual level, and by Gleason (1926) and later by Whittaker (1951, 1956, 1965) as inconsistent with the empirical data on natural plant communities. Improved data on the distribution of organisms has demonstrated that virtually all species, plants and animals, are highly clumped and patchily distributed within their overall geographic range (reviewed by Lawton et al. 1994, Brown 1995). This finding has been the empirical inspiration for the new discipline of metapopulation biology (Hanski and Gilpin 1997), which explicitly recognizes the patchy nature of most species distributions.

The best evidence of the ephemeral nature of communities comes from paleoecology. For example, the fossil pollen record from eastern North America and Europe reveals that many pre-Holocene, full glacial, and previous interglacial plant communities are very different from modern communities (Davis 1976, 1986, 1991, Overpeck et al. 1992). Overpeck et al. analyzed 11,700 fossil pollen samples and 1744 modern samples to reconstruct changes in North American vegetation over the last 18,000 years. They document a record of continuously changing plant communities with time. Before the Holocene, vegetation biomes without modern analog were widespread in response to climates and patterns of climate change that no longer exist. The spatial extent of vegetation biomes with no modern analog increases monotonically with greater time depth into the past (fig. 1.4). The data on the geographical history of the paleomigrations of individual species also support Gleason's individualistic hypothesis (Webb 1988, Huntley and Webb 1989, Davis 1991). However, there are other paleoecological studies that reveal cases of long-term community persistence in which a particular suite of species dominates, punctuated by abrupt changes from one apparently stable state to another (Jackson et al. 1996, Brett and Baird 1995). Whether climatic forcing or endogenous processes in the community cause these abrupt changes is generally not known. Whatever the cause, the evidence is strong that communities undergo profound compositional changes, sometimes gradual, sometimes episodic, on timescales of centuries to millennia and longer (e.g., Jackson et al. 1996).

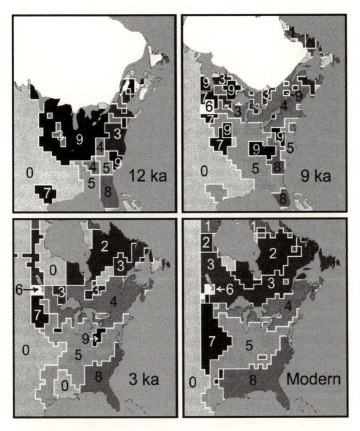

FIG. 1.4. Maps of the paleo-vegetation of eastern North America reconstructed from fossil pollen data, shown at intervals back 12,000 years bp. Mapped vegetation was averaged on a grid cell size of 200 km. The maps indicate continuous change in the distribution of vegetation biomes. Vegetation biomes having no modern analog (in black) become more widespread geographically the farther back in time one goes. Numbers indicate the following areas: 0: No data. 1: Tundra. 2: Forest tundra. 3: Boreal forest. 4: Northern mixed forest. 5: Deciduous forest. 6: Aspen parkland. 7: Prairie/grassland. 8: Southeastern forest. 9. No modern analog. After Overpeck et al. (1992).

Improved data from much deeper in the fossil record are becoming available to test ideas about the stability of fossil assemblages. It is now possible to analyze community changes over mere thousands to hundreds of thousands of years because of much more fine-scale sampling of the spatial and temporal extent of fossil assemblages. Based on detailed studies of Silurian and Devonian faunas, Brett and Baird (1995) proposed the "coordinated stasis" hypothesis (a niche-assembly hypothesis). They reported faunal compositional data on at least fourteen sequential time blocks, each lasting several million years. During these times, faunal composition and relative species abundances were relatively stable, with little species turnover due to extinction, speciation, or migration. These intervals were separated by bursts of rapid change in composition lasting less than 10% of the periods of stasis. However, Patzkowsksy and Holland (1997) have more recently challenged the generality of coordinated stasis based on more detailed temporal and spatial data for Middle Ordovician articulate brachiopod communities. Patzkowsky and Holland report high rates of taxonomic turnover (speciation and extinction rates), even during time blocks exhibiting near constancy in total faunal diversity. These high rates of origination and extinction are not consistent with the coordinated stasis hypothesis by the same statistical criteria that Brett and Baird used to test for coordinated stasis. Their data are more consistent with the view that communities undergo continuous change, comparable to the history of postglacial vegetation change in eastern North America, but on longer timescales.

Actual ecological communities are undoubtedly governed by both niche-assembly and dispersal-assembly rules, along with ecological drift, but the important question is: What is their relative quantitative importance? Falsifying the neutral, dispersal-assembly hypothesis is nontrivial. Observations of apparent long-term community persistence and resilience in the face of perturbation do not, in themselves, disprove dispersal assembly. For example, community constancy can be mistaken for stability. Constancy in local community species composition will be observed particularly among regionally abundant species. Constancy will be observed even under continual immigration and extinction if the source metacommunity is large, thereby having slow dynamics relative to the local community, due to the law of large numbers (see chapters 5 and 10). For the same reason, drifting species assemblages can also appear to be highly resilient to perturbation. Neutral communities will return to their predisturbance species composition if the disturbance is local and if migration dynamically couples the local community to the size-stabilized metacommunity having slower dynamics.

Falsifying the niche-assembly hypothesis is likewise nontrivial. It cannot be brought down solely on the basis of spatial and temporal change in ecological communities. Proponents of niche assembly have a facile reply to the paleoecologists and biogeographers. To explain most variability in natural communities and still adhere to the niche-assembly hypothesis, one need only posit the existence of sufficient environmental heterogeneity in limiting resources on the appropriate spatial and temporal scale (e.g., Tilman 1982, 1987). On the other hand, while it is not difficult to demonstrate the existence of environmental heterogeneity (Kolasa and Pickett 1991), it is considerably harder to prove that this heterogeneity is actually causing observed patchy species distributions and spatial variability among communities (Naeem and Colwell 1991). Also, by using small-scale heterogeneity as a universal explanation, proponents of niche assembly undermine their case for communities as persistent and predictable assemblages of coevolved niche specialists.

The argument between the niche-assembly and dispersal-assembly perspectives is long-standing. It has persisted so long precisely because each perspective has strong elements of truth and because reconciling them is nontrivial. This is one of the most fundamental unsolved problems in ecology today. I am convinced that a truly synthetic theory for ecology must ultimately reconcile these divergent perspectives. Applied ecology and conservation biology and policy critically depend on which perspective is closer to the truth, a fact that is not as widely appreciated as it should be. A major motivation for writing this book has been the search for the grail of reconciliation. Reconciling these perspectives is the underlying, if often unstated, theme running through this book. I believe I have made some significant theoretical progress on this question, and in chapter 10 I discuss more fully my ideas for the essential ingredients of such a reconciliation.

The organization of the book is as follows. The intellectual roots of the unified neutral theory are traced in two introductory chapters (chapters 2 and 3). The unified theory follows and is presented in two parts, divided by spatio-temporal scale. The first part considers the relatively fast dynamics of local communities or islands (chapter 4). The second part addresses the much slower dynamics of metacommunities on macroscopic spatial and temporal scales (chapter 5). These two scales are then unified into a single theory of the evolution and equilibrium maintenance of species richness and relative species abundance on continuous landscapes, which forms the basis for a dynamical theory of species-area relationships (chapter 6). Chapter 7 examines the theory in relation to metapopulations and the spatial distribution of biodiversity on the metacommunity landscape. I then explore the theory in relation to phylogeny and the implications of two modes of speciation for the evolution of metacommunity diversity (chapter 8). Chapter 9 focuses on the generality of the theory, and on sampling, parameter estimation, and testing hypotheses under the theory. In the concluding chapter (chapter 10), I revisit the themes of chapter 1 and speculate about how dispersal-assembly theory might be reconciled with niche-assembly theory and thereby lead to a truly comprehensive theory of biodiversity and biogeography in the future.

This book is by no means the first exercise of these or similar questions (e.g., Brown and Gibson 1983, Pimm 1991, Ricklefs and Schluter 1993, Huston 1994, Brown 1995, Rosenzweig 1995, Hanski and Gilpin 1997), nor will it be the last. However, the present work is unique, to my knowledge, in being the only explicit effort to construct a more general theory of biodiversity and biogeography on the original theory of island biogeography. The premise of this generalized theory is that MacArthur and Wilson were onto something important—namely, that dispersal assembly, ecological drift, and random speciation are reasonable approximations to the large-scale behavior of biodiversity in a biogeographic context. In essence, they took a statistical-mechanical approach to understanding macroecological patterns of biodiversity. I believe that this approach will prove more successful in the long run than attempts to scale up from the reductionistic approach that has preoccupied community ecology for so much of the twentieth century.

My premise in writing this book is that a good neutral theory would be enormously beneficial to the intellectual growth and maturation of ecology. In defense of the unified theory developed here, this is no mere verbal argument. The majority of results must be accorded the status of mathematical theorems because they are proofs that follow inevitably from the assumptions. Therefore, if the theory is "wrong," it will not be because the mathematics is incorrect, but because one or another crucial assumption of the theory has been violated by nature. One of the hallmarks of good theory is to fail in interesting and informative ways. Despite its simplicity, the unified theory generates a host of intriguing, nonobvious, often remarkably accurate, and, above all,

7. The niche-assembly perspective asserts that ecological communities are limited membership assemblages of species that coexist at equilibrium under strict niche partitioning of limiting resources.

8. The dispersal-assembly perspective asserts that ecological communities are open, continuously changing, nonequilibrium assemblages of species whose presence, absence, and relative abundance are governed by random speciation and dispersal, ecological drift, and extinction.

9. The argument is long-standing because both perspectives have strong elements of truth. Taking the first steps to reconcile these divergent views of ecological nature is the underlying theme of this book.

CHAPTER TWO

On Current Theories of Relative Species Abundance

No other general attribute of ecological communities besides species richness has commanded more theoretical and empirical attention than relative species abundance. Commonness, and especially rarity, have long fascinated ecologists (Rabinowitz et al. 1986, Hubbell and Foster 1986a, Gaston 1994), and species abundance is of central theoretical and practical importance in conservation biology (Soulé 1986). In particular, understanding the causes and consequences of rarity is a problem of profound significance because most species are uncommon to rare, and rare species are generally at greater risk to extinction.

Given its central importance, it is surprising that relative species abundance is missing entirely from MacArthur and Wilson's theory of island biogeography. This is all the more surprising because there is a lengthy discussion of extinction and its relationship to population size in their 1967 monograph. Just a few years earlier, MacArthur (1957, 1960) published his two famous theoretical papers on relative species abundance. These papers were steeped in niche-assembly theory and did not fully break out of the static, equilibrium mode of thinking about ecological communities. However, MacArthur's broken-stick hypothesis did raise the possibility of random community assembly by asking what relative species abundances would be if they were set by randomly apportioned limiting resources.

Before attempting to extend the theory of island biogeography, it is useful to outline the major theoretical and empirical milestones in the study of relative species abundance. More detailed reviews can be found elsewhere (Engen 1974, Pielou 1966, 1975, May 1975, Engen 1978, Gray 1987, Tokeshi 1993, 1997). Two major approaches to the study of the distribution of individuals per species have been taken: inductive and deductive. In the early years, when the study of relative species abundance was in its infancy, an inductive approach dominated. Observed distributions of the numbers of individuals per species in collections were fit to statistical distributions with little or no attempt at theoretical explanation or at defining the sampling universes from which the collections were made.

A milestone of the inductive approach was the publication of Corbet (1941) and, two years later, of Fisher, Corbet, and Williams (1943). The entomologist Steven Corbet had collected abundance data on 620 species of butterflies in Malaya. Williams (1939, 1940) had similar data on the abundance of moths collected over a four-year period at light traps at Rothamsted Experimental Station in England. When the number of species was tallied into abundance classes, i.e., species represented by a single individual, by two individuals, and so on, they noticed that the series was a relatively smooth hyperbolic progression, with many rare and few common species. Corbet and Williams took their data to Ronald Fisher, who asumed that the "true" relative abundances of species in nature would be well described by a gamma function. However, this distribution would be sampled, and Fisher assumed that the number of individuals collected of a given species would be Poisson distributed because most species were rare and represented by only a few individuals in the samples of Corbet and Williams. The resulting compound distribution was negative binomial. However, there was a problem because the zero abundance class (species too rare to be sampled) was obviously not observable, so Fisher truncated the negative binomial to eliminate the zero class. Then, having no way of estimating

31

how many species were not sampled, Fisher assumed the number of species in the community was effectively infinite. Fisher obtained a one-parameter distribution that he dubbed the logarithmic series, derived from the negative binomial as a limiting case (shape parameter set to zero).

According to the logseries, as it is now generally called, the number of species in a collection having n individuals will be given by $\alpha x^n/n$, where x is a positive constant $0 < x < 1$ and α is a measure of diversity, which in the expectation is equal to the number of singleton species divided by x. Thus, the number of species with $1, 2, 3, 4, \ldots, n$ individuals will be given by

$$\alpha x, \quad \alpha x^2/2, \quad \alpha x^3/3, \quad \alpha x^4/4, \ldots, \alpha x^n/n, \quad \text{for } 0 < x < 1.$$

Adding all terms, the total number of species, S, is expected to be $\alpha[-\ln(1-x)]$, and the total number of individuals in the collection, N, is $\alpha x/(1-x)$. The parameter α, known as Fisher's α, is a widely used measure of species diversity because it is theoretically independent of sample size (Fisher et al. 1943). However, empirically α is only approximately constant, changing slowly over large ranges in sample size (Hairston 1959, Magurran 1988, Condit et al. 1996). Fitting the logseries always results in the singleton category having the most species (fig. 2.1).

A few years later, Preston (1948) criticized the logseries on the grounds that it was not a good fit to data that he had assembled, primarily on bird species abundances. Preston argued that relative abundance distributions were more often bell-shaped curves, such that species having intermediate abundances were more frequent than very rare species. Preston also noted that the distributions were nonnormal. However, when he log transformed his species abundance data, he discovered that the relative species abundance curve could then be normalized (fig 2.2). Preston introduced a simple way to display the lognormal distribution of relative

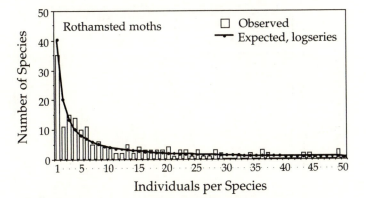

Fig. 2.1. An example of the use of the logseries distribution to fit data on species abundance in collections of months at light trap over a 4-year period at Rothamsted Field Station, U.K. The logseries always predicts that the abundance class of singleton species will be the largest class.

species abundance. He erected doubling categories of abundance (1, 2, 4, 8, et seq.), and counted the species having abundances falling in each category. Species having exactly 1, 2, 4, 8, . . . individuals were divided equally between adjacent abundance categories. He called these doubling classes "octaves" in analogy to octaves of a musical scale, which represent a doubling of the frequency of musical pitch. This classification of species into doubling abundance classes effectively log transforms the relative abundance data to the log base 2. He chose log base 2 for the simple practical expedient of spreading the distribution of species abundances over more categories to make its shape more apparent. Using any larger number for the base of log transforming the distribution would only reduce the number of categories displayed, depending on the range in relative species abundances.

The lognormal distribution is continuous, not discrete as in the case of the logseries. However, Preston's method of categorizing abundances provides a simple way to approx-

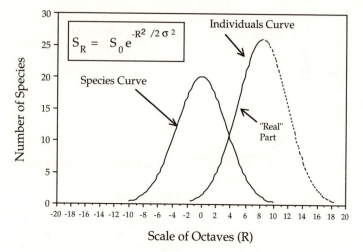

FIG. 2.2. Preston's "canonical lognormal" hypothesis. Preston argued that rare species were less numerous than species of middle abundance, and that species abundances were lognormally distributed. He counted species in doubling classes of abundance called "octaves." He further argued that there was a special relationship between the Species Curve and the Individuals Curve (discussed later), summed over all species in each octave. This canonical relationship predicted that the Individuals Curve would reach its modal value in the octave containing the most common species in the community. Redrawn from MacArthur and Wilson (1967).

imate the distribution by a discrete-valued function, as follows. Let S_0 be the number of species in the modal octave of abundance. Let S_R be the number of species in the Rth octave (or doubling abundance class) to the left or right of the modal octave. Then the so-called Species Curve can be written as

$$S_R = S_0 e^{-a^2 R^2}, \quad R = 0, 1, 2, \ldots,$$

where a is a constant that depends on the variance of the lognormal, $a = 1/\sqrt{2}\sigma$. Note that the distribution is symmetrical about the mode, located at $R = 0$. Fitting the Species Curve can be done approximately by taking natural logs, and regressing $\ln(S_R)$ on R^2, a regression having slope $-a^2$

and intercept $\ln(\mathcal{S}_0)$. More accurate fitting of the continuous lognormal distribution to the data on individual species abundances requires using a maximum likelihood technique for a truncated lognormal (Bulmer 1974, Slocumb et al. 1977). Over the past half century, the lognormal distribution has been fit successfully to a far larger number of relative species abundance distributions than has the logseries distribution, particularly as larger sample sizes have become available (Sugihara 1980).

Of almost equal importance to Preston's discovery of the widespread lognormality of the distribution of individuals per species was his recognition of the effect of sample size on the distribution. The importance of sample size had generally been overlooked because of the theoretically expected constancy of Fisher's α in collections of different sizes (Hairston 1959, Routledge 1980). Preston argued that the shape of the relative species abundance distribution observed by Fisher et al. (1943) was an artifact of small sample size. In the logseries, the expected number of species is always largest in the rarest abundance category, consisting of singleton species. However, in a small sample, one should observe only a truncated distribution of relative abundances—one comprising only the most common species. This is because common species are generally collected sooner than rare species.

As sample size increases, Preston (1948) predicted that more and more of the lognormal distribution would be revealed. He conceptualized the unveiling of the underlying lognormal distribution by the leftward movement of a "veil line" across the distribution (fig. 2.3). For any given sample size, the veil line is positioned just to the left of the rarest abundance category that is observable (i.e., species represented by a single individual). Species whose expected relative abundances are fractional for a given sample size on average are unobserved and lie to the left of the veil line because they are still too rare to have been collected

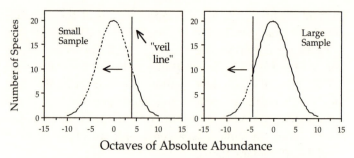

Fɪɢ. 2.3. The effect of increasing the sample size on the apparent distribution of individuals per species. As the sample size is increased, the category of rarest species, represented by singletons, moves farther and farther to the left. As more octaves of abundance are expressed in the sample, the mode is eventually seen. Thereafter, increasing sample sizes produce decreases in species number in the octaves containing the rarest species. This effect was described as an "unveiling" of the distribution, and the sampling cutoff at just below one individual per species was called the "veil line."

even once (fig. 2.3). The mode of the distribution is eventually revealed when further increases in sample size result in decreasing rather than increasing species counts in the rarest abundance category. The prediction of the existence of an interior mode to the distribution was borne out for the Rothamsted moth data after more collection years were added (Williams 1964) (fig. 2.4). Routledge (1980) would later note that the logseries distribution can easily be mistaken for the right-hand tail of a lognormal distribution.

Another useful way of graphically presenting relative species abundance data was popularized by Whittaker (1965), who named his plot the dominance-diversity curve (fig. 2.5). This curve is a graph of the logarithm of the abundance of a species on the y-axis against the rank in abundance of the species on the x-axis. Common species are assigned low ranks and appear to the left. On such a plot, the logseries appears linear, whereas the lognormal is curvilinear, at first steep over the low ranks of the commonest species, then shallower over species of middling abundance, and finally

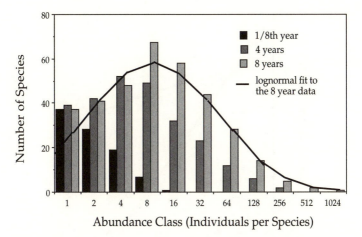

FIG. 2.4. As the survey of months at light traps at Rothamsted Field Station was extended over more years, the distribution of individuals per species became lognormal, as Preston predicted. Modified from Williams (1964).

becoming steep again over the rarest species at high ranks. Examples of such dominance-diversity curves were presented in the first figure of this book (fig. 1.1).

The logseries and lognormal were the principal "inductive" approaches to the study of relative species abundance. These were followed by theoretically inspired "deductive" approaches based on hypotheses about how ecological communities were organized.

In 1957 MacArthur published a paper in the *Proceedings of the National Academy of Sciences* critical of the inductive or "statistical" approach to studying the distribution of individuals per species in communities. This was followed by a second paper a few years later (1960). MacArthur's papers ignited a decade of great interest in deductive theoretical models of relative species abundance. He believed that the lognormal patterns of relative species abundance were so ubiquitous that there had to be an underlying general mechanism that theory could elucidate. MacArthur once confided

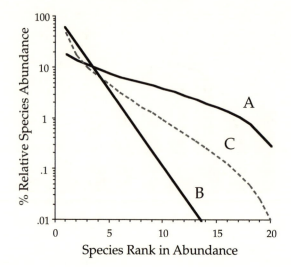

FIG. 2.5. Dominance-diversity curves, popularized by Robert Whittaker. The logarithm of percent relative species abundance is plotted on the y-axis against the rank of the species in abundance on the x-axis, with the commonest species at low rank. Curve A is the shape characteristic of MacArthur's "broken-stick" hypothesis, and is flatter than the typical lognormal observed for natural communities. Curve B is the "niche-preemption" model of Motomura, which produces a straight-line dominance-diversity curve on a semilog plot, as does the logseries distribution of Fisher et al. (1943). Curve C has the characteristic S-shape of lognormal distributions of relative species abundance. See text for further explanation. Figure redrawn from Whittaker (1975).

in me that he began by looking for a simple niche-based theory that would produce an S-shaped dominance-diversity curve like that of the lognormal. What if, MacArthur reasoned, groups of trophically similar species in ecological communities simply randomly divide up a common pool of limiting resource and their relative abundances were proportional to the fraction of total resource each utilizes? MacArthur idealized the resource pool as a stick of unit length. Suppose a community of S species randomly divides up the common resource. Now randomly partition the resource pool by throwing $S - 1$ random points onto the unit stick. Then, break the stick at each random point, and rank the fragments from shortest to longest. The expected relative abundance of the ith rarest (shortest) species, y_i, should then be given by

$$\mathrm{E}(y_i) = \left(\frac{1}{S}\right) \sum_{x=i}^{S} \left(\frac{1}{x}\right).$$

This is MacArthur's "broken-stick" hypothesis. Reasonably good fits to the broken-stick model have been found in narrowly defined communities of closely related species of birds (MacArthur 1960), minnows and ophiuroids (King 1964), and *Conus* gastropods (Kohn 1969). The distribution of fragment lengths of a randomly broken unit line segment was previously known (Whitworth 1934), but the theory of relative species abundance was a new application of the distribution. In his 1960 paper, MacArthur added embellishments to the broken-stick model, such as allowing niches to "overlap"; but the resulting relative abundance distributions agreed less well with the empirical data and were not pursued. As it turned out, the relative abundance distribution predicted by the original broken-stick model was itself not a lognormal, although it did predict an S-shaped dominance-diversity curve. The predicted distribution was often too even (fig. 2.5): common species were not common enough, and rare species were too common.

Meanwhile, Whittaker (1965) had rediscovered a much earlier deductive approach in the work of the Japanese ecologist Motomura (1932), which had been inaccessible to most Western ecologists because it was published in Japanese. Motomura studied a series of simple plant communities characterized by high dominance. He found that a steep geometrical distribution characterized relative abundances in these communities quite well. Whittaker (1965) dubbed Motomura's model the "niche-preemption" model. It was conceptually similar to the broken-stick hypothesis, except that the partition of limiting resources was nonrandom. Suppose that the community is characterized by strict hierarchical dominance. Expected relative species abundances were found by applying the following rule: Let the most dominant species sequester fraction k of the total resource pool (e.g., space for space-limited plants), leaving fraction $1 - k$ for all other species. Then let the second most dominant species sequester the same fraction k of the remaining resource, leaving fraction $(1 - k)^2$ for all remaining species, and so on. This distribution produces a straight line on a dominance-diversity plot. May (1975) suggested that this model might be one theoretical explanation for the logseries of Fisher et al. (1943), which also produces a linear dominance-diversity plot. The assumed constancy of the fraction k in the niche-preemption model has never been adequately justified, however.

These theoretical efforts still left the lognormal unexplained. This was an unsatisfactory situation because the lognormal is one of the best documented empirical generalizations in community ecology. For a time after the apparent failure of deductive approaches, the general opinion was that the lognormal was of little or no biological interest (Boswell and Patil 1981). It was repeatedly pointed out that lognormals could arise simply as the result of the multiplicative interaction of many normal random processes affecting the growth of populations (e.g., May 1975, Caswell 1976), or that lognormals could arise by combining unrelated samples (Routledge 1980).

However, interest in the lognormal was rekindled again when Sugihara (1980) argued that the lognormals that describe relative species abundance were not just any lognormals, but were a special class of so-called canonical lognormal, a term coined by Preston (1962) eighteen years earlier. By *canonical*, Preston meant that there was a special relationship between the Species Curve, discussed earlier, and the Individuals Curve for the same community. The Individuals

Curve is constructed by summing the abundances of the species in each octave of the Species Curve. Suppose there are n_0 individuals per species in the modal octave. Then the number of individuals per species in the Rth octave to the right of the mode, N_R, is $n_0 2^R$, and the number in the Rth octave left of the mode is $n_0 2^{-R}$. Thus, the Individuals Curve is given by

$$N_R = S_0 n_0 2^R e^{-a^2 R^2},$$

which can be rewritten

$$N_R = S_0 n_0 e^{(\ln 2/2a^2)^2} \cdot e^{-a^2 [R - (\ln 2/2a^2)^2]},$$

which demonstrates that the Individuals Curve is also a lognormal with the same variance but with a mode displaced by $\ln 2/2a^2$ to the right (MacArthur and Wilson 1967) (fig. 2.2). Most or all of the right half of the Individuals Curve is missing, however, because the distribution cannot extend beyond abundance octaves that still have species. It is truncated at or near its mode, which tends to be located in the octave containing the most abundant species in the community (fig. 2.2). This is not a necessary relationship, but it is empirically found in a wide variety of communities (Sugihara 1980).

With sufficiently large samples, and assuming relative species abundances are indeed lognormally distributed, essentially all species in the community will be tallied over the octaves in which $S_R \geq 1$ (or $1/2$ if one divides the count of singleton species between $S_R = 1/2$ and $S_R = 1$). The total J of all individuals in the community is $J = \sum N_R$. Taking the functional relationship between the Species Curve and the Individuals Curve as a postulate, Preston could then show that once J was specified (or more precisely J/n_r where n_r is the abundance of the rarest species, typically set to unity), then all the other parameters of the lognormal could be calculated, including the total number of

41

species, the variance of the distribution, the modal number of species, and the modal species abundance (Preston 1962). Because of this parameter interdependency, Preston called this ensemble of relationships "canonical."

The importance of the canonical relationship to Sugihara (1980) was much more than the fact that the lognormals of relative species abundance were somehow special. Sugihara revisited the broken-stick model, following a comment of Pielou (1975) that repeatedly or sequentially breaking the broken stick would eventually produce a lognormal distribution of fragment lengths. The sequential breakage procedure is as follows. Take the stick and make the first random break. Choose one of the two fragments at random and break it randomly. Then choose one of the three fragments at random and break it, and so on. What Sugihara discovered was that the resulting distribution was not only lognormal, but it was also canonical *sensu* Preston (1962). Sugihara (1980) made the analogy of sequential breakage to the action of a rock-crushing machine, which produces a lognormal distribution of particle sizes (Pielou 1975).

With the wisdom of hindsight, one can identify many shortcomings of these attempts to derive a theory of relative species abundance from first principles. Not the least of the shortcomings are the supposed first principles themselves. In MacArthur's broken-stick model, for example, it is unclear what the competitive mechanism would be that was supposed to randomly partition resources, nor is it clear what spatial and temporal scales pertain to this partitioning. Moreover, the number of species in the community is a free parameter that cannot be derived from the first principles in the theory. Expected relative species abundances are only determined once the number of species partitioning the resource base has been specified. In the niche-preemption model, the assumed perfect constancy of the fraction of remaining resources that is preempted by each succeeding species in the competitive hierarchy remains unexplained by any biological mechanism, and borders on the mystical. In the sequential breakage model, which does at least generate canonical lognormal distributions, there has not been a satisfactory biological interpretation of the rock-crushing analogy. One proposal is that sequential breakage is somehow analogous to multidimensional niche partitioning. However, the biological analog to what is done mathematically in sequential breakage is still obscure. Moreover, there is no "stopping rule" inherent in the theory that fixes how many sequential breaks to carry out. This means that the number of species in the community is once again a free parameter that does not follow from the theory. Finally, all of these models can also be faulted for having little or nothing to say about sampling issues or how they might be tested with data from real communities (Pielou 1975, Tokeshi 1993, 1997). To its credit, the statistical approach of Preston (1948, 1962), while inductive, at least was firmly grounded in a sampling theory. The greatest conceptual weakness of all these theories, however, is that they are static and do not arise in any clear way as a necessary dynamical consequence of birth, death, and dispersal processes in natural populations. Thus, it is little wonder that there is no connection between these theories and the dynamical theory of island biogeography of MacArthur and Wilson.

One might therefore have expected classical dynamical theory in community ecology to have attempted to explain relative species abundance. However, when we turn to this extensive body of theory for assistance, we find that it is almost totally silent on the subject of relative species abundance. On reflection one can understand why this is so. The edifice of theory in community ecology is constructed largely on the foundation of the classical Lotka-Volterra equations of competition (Levins 1968, Rose 1987), or on their mechanistic, resource-based counterparts (Hsu et al. 1977, Tilman 1982). In the equilibrium analysis of multispecies Lotka-Volterra competition, the focus has been almost exclusively on community stability and species coexistence, not on relative species abundance. No one to my knowledge has ever found any compelling theoretical basis for lognormal relative abundances in this body of theory. Lotka-Volterra theory is sufficiently parameter rich that no doubt someone could cleverly choose carrying capacities and competition coefficients that would yield a canonical lognormal distribution at equilibrium. A comparable exercise could be carried out in the context of resource-based competition theory. In either case, however, the choice of parameter values would be completely gratuitous. The parameter values would have no basis in theory, let alone have empirical necessity. Such an exercise would be even less justified than the theories of relative species abundance we just reviewed. For these reasons, I have little confidence that the current mainstream theoretical approaches to resource competition in communities will ever produce a successful dynamical theory of relative species abundance. At the end of the day, current niche-assembly theories of relative speices abundance fail to explain either the lognormal or the universality and invariance of Fisher's α.

In recent years, as larger sample sizes of relative species abundance have become available and the abundances of very rare species have become better known, it has become increasingly apparent that observed distributions of relative species abundance are in fact, seldom lognormally distributed. Observed distributions appear to be lognormal to the right of the mode in the right-hand tail representing common species. But they almost always show strong negative skewness, manifest as a large excess of rare and extremely rare species over that predicted by the symmetrical lognormal. This phenomenon is exemplified by the distribution of relative abundances of British breeding birds (Gibbons et al. 1993, Gregory 1994) (fig. 2.6). The right-hand tail looks like a perfectly respectable lognormal dis-

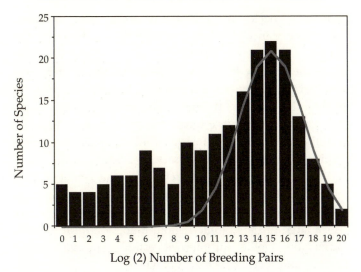

FIG. 2.6.   Negatively skewed distribution of relative species abundance for all British breeding birds. Note the poor fit of the lognormal to the left-hand tail of rare and extremely rare species. Data extracted from Gibbons et al. (1993). See also Gregory (1994).

tribution, but when the lognormal is fitted only to data from the right-hand tail, the lack of fit to the left-hand tail is immediately apparent. Anticipating a result that will be proven later in this book, the unified theory predicts the existence of a new statistical distribution of relative species abundance, called the *zero-sum multinomial*. This new distribution also exhibits negative skewness, the extent of which depends upon local community or island size and the immigration rate.

Preston (1962) entertained the hope of estimating the total number of species in a community from subsamples short of a total count. He reasoned that if distributions of relative species abundance were indeed lognormal, then one only had to collect sample sizes large enough to find the mode of the distribution. One would simply add the number of species in the modal octave to twice the number of species in higher octaves of abundance. However, the asymmetry of observed distributions of relative species abundance means that Preston's method will always underestimate the actual number of rare species in the community. As we shall demonstrate, the new zero-sum multinomial distribution will enable us to estimate the total number of species as well as their relative abundance in arbitrarily large geographic regions. All the necessary parameters of the distribution can be estimated from relatively small samples. Thus, the unified theory provides a powerful new method for estimating the total biodiversity in a given community or taxon over a large biogeographic region, filling a critical need in conservation biology for improved estimates of regional biodiversity and relative species abundance.

SUMMARY

1. The earliest theories of relative species abundance were inductive and based on fits to statistical distributions with no underlying ecological theory. Examples are Fisher's logseries and Preston's lognormal. Fisher's logseries distribution generates a diversity parameter known as Fisher's α. This parameter relates the number of species in a sample to the number of individuals in the sample. Fisher's α is widely used to characterize species diversity in collections because it is nearly invariant with increasing sample size.

2. Preston criticized Fisher's logseries because it always predicted that the rarest species would be the largest category of abundance (singletons). Preston's data produced distributions of relative species abundance with an interior mode. Preston argued that Fisher's logseries was an artifact of small sample size, which has generally proven in later years to be correct. However, the log-

normal left unexplained the apparent universality and invariance of Fisher's α.

3. Later theories of relative species abundance were deductive and based on one or another ecological theory of community organization. Each of these theories, such as MacArthur's broken-stick hypothesis, proposed simple a priori rules for how limiting resources would be apportioned among competing species in a community.

4. These deductive theories of relative species abundance can be faulted with the wisdom of hindsight. Their most serious problem is that they do not derive in any straightforward way from fundamental birth-death-migration processes in population dynamics. Also, the number of species in the community is a free parameter that cannot be predicted from first principles in any of the deductive theories.

5. Newer and much larger data sets on relative species abundance indicate that virtually all distributions have long and negatively skewed tails of very rare species. None of the current theories of relative species abundance, including Preston's lognormal, satisfactorily explain this negative skewing. This skewness means that the lognormal will always underestimate the number of rare species. It also means that relative abundance distributions are not canonical in the sense that there is no special relationship between the Individuals Curve and the Species Curve, as postulated by Preston.

6. The unified theory predicts the existence of a new distribution of relative species abundance called the *zero-sum multinomial*, which exhibits negative skewness, the extent of which depends on island size and the immigration rate.

Dynamical Models of the Relative Abundance of Species

In the previous chapter, I temporarily set aside a small and curious set of dynamical models of communities, most of which have not received the attention they deserve from mainstream theoretical community ecology. I have chosen to discuss them separately because I believe they are closer to the right track for developing a successful dynamical theory of biodiversity and relative species abundance. These models differ from the models discussed in the last chapter in that they explicitly incorporate the demographic processes of birth, death and dispersal.

In the mid-1970s, when most eyes were still focused on the classical, niche-based theory of community ecology, Caswell (1976) made a bold attempt to create a neutral theory of community organization. Borrowing mathematical machinery from the theory of neutral evolution in population genetics, Caswell erected three models, only the first of which will be discussed here. In model I, communities are essentially collections of completely noninteracting species in which each species undergoes an independent random walk in abundance. Therefore, the total size of the community fluctuates. New species enter the community as a Poisson process (i.e., a rare event) with probability ν per unit time. This immigration probability, as in the theory of island biogeography, is independent of the identity of the species and of the number and identities of the species already present, except that only species not currently present are allowed to immigrate. This is equivalent to assuming that immigration makes a negligible contribution to the population dynamics of a species already present. Each new immigrant species becomes the founder of a line of descendants. Caswell assumed a linear birth-death process in which the stochastic per capita birth and death rates, λ and μ, are assumed to be equal, corresponding to the deterministic case of an intrinsic rate of increase, r, of zero. In other words, each species population is as likely to increase as it is to decrease per unit time. This is a pure drift process or random walk. The transition probabilities from a population of size N_i to size $N_i - 1$, N_i, or $N_i + 1$ at time $t + dt$ are linear functions N_i of at time t, as follows:

$$\Pr\{N_i - 1|N_i\} = \mu N_i$$

$$\Pr\{N_i|N_i\} = 1 - (\lambda + \mu)N_i$$

$$\Pr\{N_i + 1|N_i\} = \lambda N_i.$$

Note in this model that λ and μ must be chosen to be sufficiently small that the expression $(\lambda + \mu)N_t < 1$.

Caswell's models II and III are similar to model I except that, instead of a constantly fluctuating community size, community size is held constant. The addition of constancy in community size is crucially important, and its absence is one of the major weaknesses of Caswell's model I. Unlike the neutralists in population genetics, Caswell did not defend his neutral model as a realistic description of actual community dynamics. Caswell's purpose in creating these models was to provide a neutral benchmark for comparison with the structure and dynamics of actual ecological communities. He developed a series of deviation statistics to measure departures of real communities from the predictions of his neutral models.

In any case, Caswell's neutral results differ substantially from observed community relative abundance patterns.

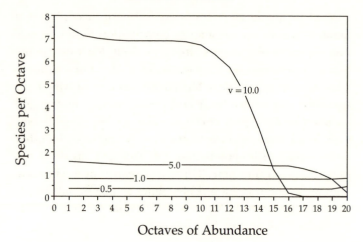

FIG. 3.1. Relative abundances predicted by Caswell's neutral models are decidedly not lognornal-like. The family of curves in the figure is drawn for various rates of new species addition to the community (parameter ν) per unit time. The distributions are extremely flat across many octaves fo abundance. Only when the immigration rate is very high does the distribution develop a right-hand tail. Moreover, these distributions are transient because the number of species in the models increases with time without bound.

The relative abundance distribution predicted by Caswell's model I is decidedly not lognormal on a Preston plot of octaves of abundance (fig. 3.1). Regardless of parameter values, the distributions tend to be nearly log uniform across many octaves of abundance. Only when the immigration rate was extremely high was there a even a hint of a lognormal like right-hand tail for the commonest species. The distributions also give no indication of having an interior mode at intermediate abundances. In fact, the distributions are closer to the logseries of Fisher et al. (1943) than they are to the lognormal (Caswell 1976), but the logseries is not a particularly good fit, either.

However, there are much more serious problems with Caswell's model. One is that the size of the community grows without bound over time. Community size, J, where J is the total number of individuals in the community, is a negative binomial random variable with mean $E\{J\} = t \to \infty$ (elapsed time), and variance as $Var\{J\} = t(t + 1) \to \infty$ as $t \to \infty$. A second major problem is that the expected number of species in the community, $E\{\mathcal{S}\}$, is linearly proportional to the colonization rate of new species per unit time, ν, and the log of elapsed time:

$$E\{\mathcal{S}\} = Var\{\mathcal{S}\} = \nu \cdot \ln(t + 1).$$

I think it is safe to assume that no one would accept these results as reasonable for real ecological communities.

Given the unreasonableness of Caswell's model, it is logical to ask, why pay so much attention to it? To my knowledge, Caswell was the first person to recognize the importance of basing a model of relative species abundance explicitly on birth, death, and dispersal processes. Moreover, as will shortly become apparent, with the addition of the biologically reasonable assumption of a finite community size and minor changes in the birth, death, and dispersal processes, a much better model is obtained. Indeed, it is unclear why Caswell's models II and III, which had a constant community size, didn't perform better than his model I. Caswell (1976) asserts that the results from all three models were qualitatively similar, but he did not actually report on the behavior of models II and III in his paper.

The importance of a finite community size cannot be overemphasized. Ecologists who work on space-limited communities seem generally more aware of this fact than others. For example, plant ecologists who study sessile plants, and marine ecologists who study intertidal or benthic invertebrate communities, recognize that there is an unavoidable physical constraint on the total number of individuals that can be packed into a given space (Hughes 1984, 1986, Yodzis 1986, Weiner 1985, Jackson et al. 1996, Harper 1977).

Space per se of course is not a resource, but it is a good surrogate variable for limiting resources that are distributed uniformly over the two-dimensional landscape, such as sunlight or planktonic food. When access to resources requires controlling a unit of space, it is reasonable to refer to space as a limiting resource. It should be noted, however, that even when space is not the limiting factor, limiting resource availability per unit area will ultimately impose a finite limit on the density of competing organisms within a given ecological community in a defined space (Brown 1995). Across communities of trophically dissimilar species, densities of organisms per unit area will vary, depending on the relative sizes and energy demands of species in the different communities. However, *within* a given community of trophically similar, competing species, the numbers of organisms per unit area should not vary too widely (i.e., by orders of magnitude) from one locale to the next.

Preston (1948) was fully aware of this fact, as were MacArthur and Wilson (1967), who noted that the total number of individuals in a defined taxon or community, J, increases linearly with the area, A, inventoried:

$$J = \rho A,$$

where ρ is the density of individual organisms per unit area. Figure 3.2 illustrates this relationship for an enumeration of trees in a closed-canopy tropical forest on Barro Colorado Island (BCI), Panama, in a 50 ha plot of mixed, species-rich, old-growth forest. The plot has been completely censused five times for all free-standing woody stems >1 cm diameter at breast height (dbh) (Hubbell and Foster 1983, 1990). This relationship holds very precisely ($r^2 \approx 1.000$) over more than five orders of magnitude of variation in area, from 1 m^2 to $5 \cdot 10^5$ m^2 (the largest area censused). It holds despite the fact that more than three hundred species occur in the BCI plot, that the species composition of the BCI plot varies

One could imagine all sorts of possible causes of variation in the density of individuals. A failure to obey this principle would suggest at least three possible conclusions. The disturbance regime could be so severe on landscape scales that the region is not, in fact, saturated, and permanent open space or unused limiting resource exists. Or there could be variation over the landscape in the overall regional supply rates of limiting resources for the community as a whole (landscape variation in potential productivity). Or one might be attempting to aggregate taxa that are trophically too dissimilar to be logically treated as members of the same metacommunity (chapter 1). Or all three possibilities could be true simultaneously. The remarkable fact is, however, that a linear relationship between the number of individuals in a well-characterized community and area is almost universally observed, at least on relatively homogeneous landscapes.

The implications of this simple—indeed, seemingly trivial—relationship between individuals and area are far more profound than at first apparent. At least two major theorems about biogeography and relative species abundance follow from this first principle. The first theorem follows immediately, namely, that *the dynamics of ecological communities are a zero-sum game.* If, as the relationship implies, the density of individuals ρ is a constant, then any increase in one species must be accompanied by a matching decrease in the collective number of all other species in the community. The sum of all *changes* in abundance is always zero. Thus, the total number of individuals in the community behaves like a conserved quantity, except that individuals of one species cannot be transmuted into individuals of another species (speciation excepted). Unlike subatomic particles, however, individuals can reproduce themselves. This reproductive property fuels a birth process that replaces individuals that die in the community in a zero-sum game. When one species is successful at reproducing beyond self-replacement, other species must have compensatory failures in reproduction.

No new individuals can be added to an ecological landscape by birth or immigration until vacancies have been created by deaths. Note that the zero-sum game does not actually require that the carrying capacity of the landscape be constant through time or space; it only requires that the landscape be biotically saturated at all times, i.e., that the biota fully track changes in limiting resources.

The second important theorem that follows from the biotic saturation of landscapes and the zero-sum dynamics of ecological communities is not immediately obvious, however, and is essentially the subject of the remainder of this book. This theorem concerns the consequences of zero-sum dynamics for the equilibrium distribution of relative species abundance in local communities and in the metacommunity, given certain rules of the zero-sum game. This theorem is the subject of formal proofs in chapters 4 and 5, but I anticipate some of the results qualitatively in this chapter.

The rules of the zero-sum game now have to be considered. How the game is played could be anything—so long as the sum of all abundance changes is zero. For example, rare species might have a per capita competitive advantage (frequency dependence), and/or per capita population growth rates may decline with increasing population size (density dependence). The question now arises, what rules do zero-sum community dynamics have to obey to result in a lognormal-like distribution of relative species abundances? The answer to this question is the second theorem to follow from our first principle, and will only finally and fully be answered in the next two chapters.

The next steps in answering this question were taken by me (Hubbell 1979), working on closed-canopy, tropical forests, and by Hughes (1984), working on benthic marine invertebrate systems, both space-limited communities. Hughes and I independently developed stochastic models of relative species abundance that were similar in certain important respects. Both models explicitly assumed zero-sum community dynamics, and both specifically modeled birth and death processes. Because Hughes's model is considerably more complex and specific than mine, however, I will not discuss his model further.

I began with the simplest possible assumption of a community obeying zero-sum dynamical rules that were neither frequency nor density dependent (except for fixed community size). Consider a model community that has J total individuals, regardless of species, so that our first principle applies and the community obeys zero-sum dynamics. Each individual occupies one space or unit of limiting resources and resists displacement by any other individual. Eventually, however, the individual dies, with probability μ per unit time, and it is replaced by a "birth." Now suppose that the replacing species is randomly drawn from the community. Let the probability that the replacing individual is of species i be given by the current relative abundance of species i. Let the current abundance of species i be N_i. Then, the transition probabilities that species i will decrease by one individual remain unchanged in abundance, or increase by one individual during one time step are given by

$$\Pr\{N_i - 1 | N_i\} = \mu\left(\frac{N_i}{J}\right)\left(\frac{J - N_i}{J - 1}\right) = \mu N_i(J - N_i)/J(J - 1)$$

$$\Pr\{N_i | N_i\} = 1 - \Pr\{N_i - 1 | N_i\} - \Pr\{N_i + 1 | N_i\}$$

$$= 1 - 2\mu N_i(J - N_i)/J(J - 1)$$

$$\Pr\{N_i + 1 | N_i\} = \mu\left(\frac{J - N_i}{J}\right)\left(\frac{N_i}{J - 1}\right)$$

$$= \mu N_i(J - N_i)/J(J - 1).$$

Thus, the probability that species i will increase by one individual is the probability that a death occurs in a species other than species i, or $\mu(J - N_i)/J$ times the probability that the next birth occurs in species i, or $N_i/(J - 1)$. Note that the probabilities that species i will increase or decrease are identical. This model is even simpler if we scale time so that a single time step is the mean time required for one death to occur ($\mu = 1$).

Species in this simple neutral model are identical and equal competitors on a per capita basis. They have identical per capita chances of dying and of reproducing. Each species has an average stochastic rate of increase, r, of zero. The dynamics of the model community is therefore a random walk. For this reason, I call this process *ecological drift* in analogy with genetic drift. Unlike Caswell's (1976) neutral model I, however, this random walk is not completely free and unfettered. The random walk is constrained by the fact that all species abundances must sum to a constant J, i.e., the sum of all positive and negative changes in abundance must sum to zero. J can be large or small depending on the density of individuals, ρ, and the size of the area occupied by the community. No species in the community can increase in abundance above the absolute maximum imposed by J, which corresponds to complete dominance. Conversely, a species can decline to zero abundance, an absorbing state that corresponds to local extinction. However, the ecological drift of a species to extinction can be quite a long process, as will be discussed in chapter 4.

I return now to the central question, namely, what rules must the zero-sum dynamics of ecological drift obey in order to obtain a lognormal-like distribution of relative species abundances? It turns out that the rules of the zero-sum game are indeed critical to the answer. As will be fully explored in the next two chapters, the answer is this: *Relative species abundances are lognormal-like if community dynamics obey a zero-sum random drift process.* This is the second theorem that follows from our first principle under zero-sum ecological drift. An even more remarkable result, which presently is still a simulation-based mathematical conjecture, is that lognormal-like distributions of relative species abundance are *not* obtained when the rules of the zero-sum game are density or frequency dependent. I will discuss the evidence for this conjecture shortly.

I use the term *lognormal-like* because the theoretical distribution of relative species abundance predicted to occur in local communities by the unified theory (chapter 5) is not a lognormal (Hubbell 1995). I have named this new statistical distribution the *zero-sum multinomial* (Hubbell 1997). The zero-sum multinomial distribution is discrete, not continuous. In many cases, however, casual inspection will not be able to distinguish a zero-sum multinomial from a lognormal when the latter is approximated by a discrete-valued function (e.g., a Preston curve). It also differs from a lognormal in having a long, attenuated tail of rare species (see chapter 5), so that the distribution is asymmetrical about the modal octave. Otherwise, the distributions are very similar, particularly in octaves to the right of the mode. Bell (2000) used simulations to explore some properties of the distribution as community size and immigration rate are varied, a topic to which I shall return in chapter 5. The distribution has similar sampling properties to Preston's lognormal distribution in that, as sample size is increased, more and more of the zero-sum multinomial is revealed. Only when sample sizes are large will the differences between the zero-sum multinomial and the lognormal become clear in the long tail of very rare species. Also, under one of the two modes of speciation studied in this book, a different distribution of relative species abundance is predicted for the metacommunity than for the local community, as we shall see in chapter 8.

It is easy to understand qualitatively why relative abundances would tend to be lognormal-like rather than normal-like under neutral zero-sum ecological drift. In the drifting community, there are many approximately normally distributed fluctuations of species about their current respective abundances. But common species will tend to fluctuate more in absolute abundance because they undergo absolutely more births and deaths per unit time than rare species. This will be true even though on a per capita basis, birth and death probabilities are the same in all species. However, the critical factor is that these species fluctuations collectively must obey the zero-sum rule. Having zero-sum ecological drift in essence multiplicatively couples the fluctuations among species. This coupling produces the lognormal-like distribution of relative species abundance observed when species frequency is plotted against arithmetic individuals per species.

The dependence of lognormal-like relative species abundances on zero-sum ecological drift is illustrated by the contrasting relative abundance patterns of tree species in three tropical forests. Each of the communities were sampled by 50 ha permanent plots. Two of the communities, Pasoh and BCI, are closed-canopy forests in which zero-sum dynamics operate. The relative species abundance distributions for Pasoh (fig. 3.3) and BCI (fig. 3.4) are lognormal-like. They are remarkably similar in their variance and in their modal octave of species abundances, despite the fact that they have very different taxonomic compositions and evolutionary histories. The main difference is that the Pasoh plot has about two and a half times as many species as the BCI plot. The third tree community is located in Mudumalai Game Reserve in the Western Ghats of southern India. Mudumalai is an open-canopied forest with a grass-dominated understory and less than 25% total tree cover (R. Sukumar, pers. comm.). In contrast to Pasoh and BCI, the open-canopied forest at Mudumalai has a distribution of relative species abundance that is decidedly not lognormal-like (fig. 3.5). The unified theory explains this observation by the fact that the canopy of the Mudumalai forest is not saturated with trees. Therefore, the population dynamics of individual tree species are to a large extent independent of one another, and are not constrained to follow zero-sum dynamics like

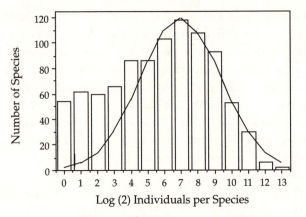

Log (2) Individuals per Species

FIG. 3.3. Lognormal-like distribution of the relative abundances of the tree species in the 50 ha plot in a closed-canopy forest in Pasoh Forest Reserve, Peninsular Malaysia. The best-fit lognormal is superimposed on a Preston plot of species frequencies per octave of abundance. Compare with the plot for the 50 ha plot on Barro Colorado Island, Panama (fig. 3.4). Note the poor fit to the rare species. This deviation in having too many rare species is another almost universal pattern in closed-canopy forests, about which more will be said later (see also fig. 3.4). Data from Manokaran et al. (1993). The graph represents counts of all stems >1 cm dbh. The zero-sum multinomial will be fit to the full curve in chapter 5.

populations of tree species in the closed-canopy Pasoh and BCI forests.

Recall that Preston (1962) claimed to have found a canonical relationship among the parameters of the lognormals that seemed to characterize many ecological communities. If J/n_r—the ratio of all individuals to the number of individuals of the rarest species in the community—was specified, then Preston argued that all of the other parameters of the lognormal could be deduced. Note that J/n_r can be rewritten in terms of area as $\rho A/n_r$. In finite samples, n_r will usually be unity, which is the smallest observable abundance. Therefore, if Preston's empirical generalization were correct, then the canonical lognormal and the relaive

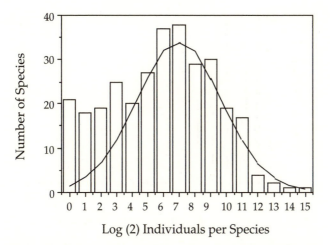

Log (2) Individuals per Species

FIG. 3.4. Lognormal-like distribution of the relative abundances of tree species in the 50 ha plot in a closed-canopy forest on Barro Colorado Island, Panama. The best-fit lognormal is superimposed on a Preston plot of species frequencies per octave of abundance. Compare with the plot for the 50 ha plot in Pasoh Forest Reserve, Peninsular Malaysia (fig. 3.3). Note the poor fit to the rare species. This deviation in having too many rare species is another almost universal pattern in closed-canopy forests, about which more will be said later chapter 5 (see also fig. 3.3). The graph represents counts of all stems >1 cm dbh. The zero-sum multinomial will be fit to the full curve in chapter 5.

species abundance distribution itself should be completely characterizable from the mean density of individuals per unit area, and the total area sampled. This is clearly false. MacArthur and Wilson (1967) did not remark on this prediction, but they did argue that the canonical lognormal predicted a quantitative species-area relationship, specifically a log-log linear relationship between the log of the number of species S and the log of J/n_r. At intermediate regional spatial scales, this is true (May 1975), but, as I shall show in chapter 6, it is decidedly not true on local spatial scales. I shall also show that the metacommunity distribution of relative species abundance cannot be completely determined by

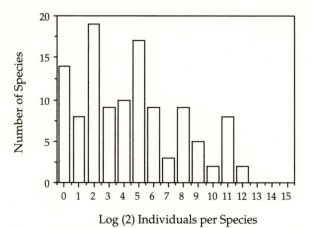

FIG. 3.5. Nonlognormal-like distributions of the relative abundances of tree species in the 50 ha plot in open-canopy woodland in Mudumalai Game Reserve, Western Ghats, India. This is a forest that has a thick grass understory that burns in most years. The forest is also disturbed by elephants. The graph represents counts of all stems >1 cm dbh. Data courtesy of R. Sukumar.

the mean density of individuals and the total area sampled. The critical missing parametric ingredients are the rates of speciation and dispersal (see chapter 6).

The relative abundance distributions predicted by the zero-sum multinomial are only accidentally and occasionally canonical *sensu* Preston (1962). This is perfectly all right, in my opinion, because it is easy to show that the canonical relationship is neither a mathematical nor an empirical necessity, and indeed that it cannot be generally true. Canonical lognormals have typically been observed in communities having relatively small numbers of species and high variance in relative species abundance, or in cases of small sample sizes that have not unveiled the rarest species. Most of Sugihara's (1980) example communities that exhibited canonical distributions were relatively species poor. On the other hand, species-rich communities in which the commonest species comprise only a small percentage of total individuals often do not exhibit canonical lognormals. For example, the lognormal-like distributions of relative tree species abundance in the Pasoh and BCI forests are not canonical. These distributions exhibit large excesses of rare species over the number predicted by the lognormal (figs. 3.3 and 3.4). Moreover, the modes of the Individuals Curves for Pasoh or BCI are not located in the ultimate octave of the Species Curve.

Preston (1980) himself later acknowledged finding communities with noncanonical lognormal relative species abundance patterns. In fact, it has to be the case that all "true" distributions of relative species abundance in nature are not canonical, a fact that will be revealed if and when they are adequately sampled. If relative abundance lognormals were always canonical, then from the logic of the putatively canonical species-area relationship given above, there should be one and only one possible number of species S in communities of density ρ and area A. In fact, the number of species in real communities occupying a fixed area is observed to vary considerably, from monodominant communities to species-rich communities—even though the total number of individuals in the fixed area does not vary greatly. Factors affecting S in a particular area (e.g., an island) include birth, death, and immigration rates—among the basic ingredients of MacArthur and Wilson's theory of island biogeography.

The breakdown of the canonical relationship occurs primarily because of the long asymmetrical tail of rare species in real communities. Suppose species relative abundances were truly lognormal and the number of species, S, in the community were fixed as Preston claims. Then, in principle, one should be able to increase sample sizes sufficiently that the last and rarest species would have abundances greater than unity. This is because, in Preston's universe, the number of species is finite, so that if a sufficiently large sample is taken, they will all be found. In order for the lognormal to be canonical, the number of octaves separating the rarest species from the commonest species must remain constant. Therefore, as the common species become commoner with increased sample size, the rarest species must also become commoner. However, in real samples, the abundance of the rarest species sampled, n_r, almost invariably remains "locked" at unity. Thus, there is no fixed relationship between the total number of individuals in the community, J, and the abundance of the rarest species, n_r. As sample size increases, new rare species never before sampled are added ever more slowly, and the rarest species become ever rarer relative to common species in a seemingly endless regression. This means that the number of octaves of abundance separating the rarest species from the commonest species grows steadily greater. This sampling phenomenon is observed and is predicted by the present unified theory, but not by Preston's canonical lognormal hypothesis.

In reference to immigration, in island biogeography theory the number of species at equilibrium on an island is sustained only under a persistent rain of immigrants. There are parallels in the theory of ecological drift. Under drift without immigration, species will gradually be lost from the community because of local extinction. However, because the landscape remains saturated with individuals, the mean abundance of the surviving species must thereby increase. The elimination of species under ecological drift can be very slow, especially for large communities (chapter 4). Without immigration, the community eventually collapses to a single species in a final equilibrium. However, with immigration (or speciation), a persistent multispecies equilibrium community is achieved with a distribution of relative species abundances that is determined by community size J and the probability of immigration into the community. In these multispecies, multinomial equilibria, the number of species is not a free parameter as it was in the static, niche-assembly theories of relative species abundance discussed earlier (chapter 2). Instead, both the equilibrium number of species as well as their equilibrium relative abundances are predictions of the unified theory.

The next dynamical model I wish to consider is quite famous. It is the lottery model of Chesson and Warner (1981) and Chesson (1986). This model of community dynamics differs from the ecological drift model in having a frequency-dependent birth process. Rare species in Chesson and Warner's model community enjoy a per capita advantage in reproduction over common species. The lottery model, originally proposed by marine ecologist Peter Sale (1977, 1980) for space-limited communities of coral reef fish, assumes that space is allocated at random or by "lottery," and it is one specific case of what Chesson and Warner called the "storage hypothesis." Because of recruitment fluctuations, rare species accrue an advantage in winning vacated space.

Chesson and Warner's model of community dynamics is as follows. Let $N_i(t)$ be the population of adults of the ith species at time t. Then the number of the ith species present one time step later is given by

$$N_i(t + 1) = (1 - \mu_i)N_i(t) + R_i(t)N_i(t),$$

where μ_i is the adult death rate and $R_i(t)$ is the time-varying, per capita recruitment rate of new adults into the population. Chesson and Warner suggest that a great number of varying environmental factors could influence $R_i(t)$. They called the model the "storage hypothesis" because recruitment fluctuations only promote frequency-dependent coexistence if adults live for more than a single reproductive season, i.e., there are overlapping generations. The average lifespan is the inverse of the adult mortality rate, or μ_i^{-1}. Therefore, there are overlapping generations if $\mu_i < 1$.

In the lottery case, the relative recruitment rate for species i at time t is simply the number of spaces vacated by deaths at time t multiplied by the fraction of all births in the community that are of species i. The total number of deaths in the community at time t is $\sum_k \mu_k N_k$. If $\lambda_i(t)$ is the per capita birth rate of species i at time t, then $R_i(t)$ is

$$R_i(t) = \sum_k \mu_k N_k(t) \cdot \left[\lambda_i(t) N_i(t) \Big/ \sum_k \lambda_k(t) N_k(t) \right].$$

Note that if in a given time period species i is the only species reproducing, then $\lambda_i(t) N_i(t) / \sum_k \lambda_k(t) N_k(t)$ is unity, and all the vacant sites are won by species i in that time period. Note also that this birth or recruitment function is identical to the drift model that forms the basis of the unified theory if the λ's are constant and unity. In fact, except for rare species recruitment advantage, the lottery version of Chesson and Warner's model is virtually identical to ecological drift. It differs in having a deterministic death process; but most importantly, it has the same zero-sum dynamics. Species are identical competitors in the following sense: if any species becomes rare, that species will enjoy the same frequency-dependent reproductive advantage as any other equally rare species, and all suffer the same frequency-dependent disadvantage if they become common.

The result that made Chesson and Warner's model famous was their proof that variability per se could promote coexistence. Prior to their analysis, the conventional wisdom was that variability was just noise that would have no effect on, or even reduce the possibility of, coexistence. For example, Turelli (1978a,b) and Turelli and Gilpin (1980) studied Lotka-Volterra competition equations with stochastic variation in carrying capacities, competition coefficients, and intrinsic rates of increase, but found essentially no effect on coexistence. However, Chesson and Warner showed that temporal variation in the birth process, but *not* in the death process, could allow two competitors, or n competitors, to coexist under conditions that would otherwise lead to competitive exclusion under constant recruitment.

A simple hypothetical example told to me by Bob Warner illustrates the basic principle. Imagine a model reef fish community of two interspecifically territorial fish. One species is common and its adults currently occupy 90 of the 100 territories on the reef. The other species is rare and its adults occupy the remaining 10 territories. Now suppose that there is 10% annual adult mortality, and it is random across the two species. Then nine deaths are expected in the common species, but only one death in the rare species. If each species reproduces annually in proportion to their adult abundance, then the only changes in abundance that will occur will be due to random drift. However, if there is a tendency for the rare species to reproduce in years when the common species does not, then coexistence will occur. This is because of very strong rare-species advantage. In this numerical example, if the rare species occupies all of the vacated territories in a given year, its population will increase by 90%, whereas the common species will have decreased by only 10%. Chesson and Warner (1981) prove that all that is required for coexistence is that there be overlapping generations and *some* temporal variability in recruitment rate; recruitment does not have to be completely asynchronous.

Chesson (1986) noted that the frequency dependence in the model is strong enough to bound each species away from zero abundance so that coexistence will occur. However, one might argue that the frequency dependence is overly and unrealistically strong. For example, according to their model, a very rare species reduced to a single individual in a very large community can potentially produce enough offspring to win every vacant site. The larger the community, the more sites that are vacated per unit time, and the rarer a species is, the stronger the rebound-from-rarity effect becomes in their model. There is essentially no upper bound to the strength of frequency dependence in Chesson and

Warner's model. In actual communities, on the other hand, dispersal limitation (Tilman 1994, Hurtt and Pacala 1995) and biological constraints on individual fecundity will set finite limits on maximum per capita recruitment success. I suspect that the realized recruitment success of most rare species will be far lower than in Chesson and Warner's model. Whatever frequency dependence rare species enjoy will often not successfully bound rare species away from zero abundance. And this must be so, because otherwise no species would ever go extinct, locally or globally.

More germane to the present discussion, however, are the implications of Chesson and Warner's model for relative species abundance. It is surprising, but this question has apparently never been asked. Without waiting for analytical results, which may be very difficult to obtain in any event, one can easily simulate the dynamics of a model community with frequency dependence in the recruitment process. I simulated tree communities of varying numbers of masting tree species. Tree species that mast—for example, the oaks in eastern North America, or the dipterocarps of Southeast Asia—are species that flower and fruit on variable intervals of several years (Curran et al. 1999). In many masting species, there is interspecific synchrony in masting years, presumably to reduce collective seed and seedling predation (Janzen 1974). In my simulations I modeled forests in which masting behavior was simply stochastic, occurring with probability ξ per year. The results were that, regardless of the value of ξ or the size of the mast, or the number of species in the community, the outcome was always coexistence, just as predicted by Chesson and Warner. However, the stochastic equilibrium relative species abundances were *never* lognormal-like. A typical example of the distribution of individuals per species from one of the simulations is presented in figure 3.6. In this figure, the abundance classes are plotted arithmetically, not log transformed, so it is easy to

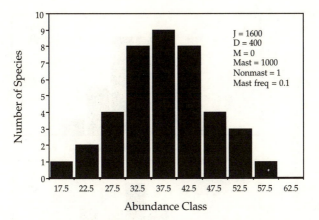

FIG. 3.6. Plot of the distribution of relative species abundance predicted by the lottery model of Chesson and Warner. Note that the histogram is of arithmetic abundance, not logarithmic abundance. The non-lognormality of the curve is apparent.

see that relative species abundances are very close to being perfectly normally distributed.

I have explored a number of other models of density and frequency dependence in addition to the lottery model, and the result has always been qualitatively the same under zero-sum ecological drift. The distribution of the number of individuals per species is never lognormal (or zero-sum multinomial). For example, one can study community dynamics and relative species abundance under stochastic logistic growth. Consider a case in which all species had a maximum carrying capacity of J (community size). Let the per capita birth and death rates, λ and μ, be the standard logistic expressions $\lambda(N_i) = 1 - bN_i$ and $\mu(N_i) = dN_i$, where slopes b and d are chosen so that $J = (b + d)^{-1}$. The probability that species i will increase by one individual in one time step is therefore $\lambda N_i / \sum_k \lambda(N_k)$, and the probability that species i will decrease by one individual is $\mu N_i / \sum_k \mu(N_k)$.

69

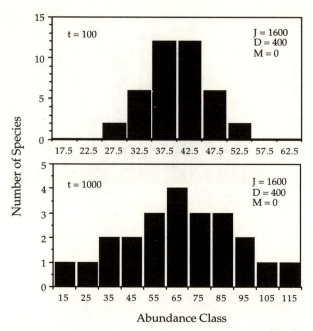

F<small>IG</small>. 3.7. Nearly normal transient distributions of relative species abundance for a community obeying zero-sum dynamics, in which each species grows according to a stochastic logistic, each with an identical carrying capacity of the community size, J. Species are only very slowly lost to extinction because of the density dependence. Note that the abundance classes are arithmetic, not logarithmic. Numbers of species represent the mean number of species in abundance classes in one hundred runs of the model.

I illustrate the qualitative behavior of the stochastic logistic community model with the results of one simulation (fig. 3.7). In this case the community size J was set at 1600 with initial conditions of forty equally abundant species having forty individuals apiece. In each disturbance cycle, a quarter of all individuals died and were replaced according to the stochastic logistic. A quasi-equilibrium, near-normal distribution of relative species abundance is achieved quite rapidly. Figure 3.7, top panel, shows the abundance distribution after one hundred disturbances. It is a quasi-equilibrium because the community loses species to extinction very slowly. As species are gradually lost, the variance in relative species abundance increases but the distribution remains very nearly normally distributed. After one thousand disturbances, the distribution is still essentially normal, not lognormal-like (fig. 3.7, bottom).

In my simulations, the distributions of relative species abundance produced by these models are always closer to being normal than lognormal. On reflection, it is clear why this is so under zero-sum ecological drift. Frequency and density dependence that give rare species a competitive advantage will tend to make rare species relatively more common, and simultaneously make common species rarer. Frequency dependence compresses the tails of the distribution of species abundances, lessening the skewness of the distribution when it is plotted arithmetically. Density and frequency dependence act as centrally tending forces that reduce the variance in abundances of species in the community by driving them toward middling abundances. In contrast, under zero-sum, ecological drift, there are no such centrally tending forces. These results do not imply the converse, however, that in the absence of zero-sum ecological drift, one cannot obtain a lognormal-like distribution of relative species abundance. The last dynamical model I discuss makes this point.

The final model represents recent theoretical work by Engen and Lande (1996). Their results are important for several reasons in the present context. First, they show that one cannot conclude that density dependence is absent from a community simply by observing relative abundance distributions that are lognormal. Second, their model allows for both demographic and environmental stochasticity in the population growth rate. Thus, Engen and Lande's model is in many ways very close to the present theory. They introduce a new class of stochastic species abundance models that includes a process of speciation and density dependence. Their paper is rather technical, however, and so only a general overview of their results are discussed here. For more detail, the reader should consult their paper.

Engen and Lande assume that new species originate slowly by a Poisson process that can be inhomogeneous in the sense that species do not have to have the same speciation rate. They study the stochastic changes in abundance of an arbitrary species by assuming that small changes can be modeled by a diffusion approximation. Let the current abundance of such a species be x, and let the stochastic distribution of abundance be $\lambda(x)$. For simplicity, assume that the speciation rate is constant (ω_0). They then show that the mean amount of time that a species undergoing stochastic population fluctuations spends at abundances between x and $x + \partial x$ is given by

$$\lambda(x) = 2\omega_0 \frac{1}{v(x)} \exp\left[\int_1^x \frac{2m(u)}{v(u)} du\right],$$

where $m(x)$ and $v(x)$ are the infinitesimal mean and variance of abundance. Now, let us suppose that population growth is density dependent. Consider the following general differential equation for density-dependent growth:

$$\frac{dx}{dt} = rx - xg(x),$$

where r is the intrinsic rate of increase of the population. The density-dependent function $g(x)$ can be anything, but a well-known function is the Gompertz, where $g(x) = \gamma \log(x)$ (they actually use a form of density dependence that is very close to but not exactly the Gompertz in order to get a closed integral solution). The per capita growth rate of the populations falls off linearly with the logarithm of abundance. Now, to make the growth equation stochastic, let the intrinsic rate of increase r have both an environmental variance σ_e^2 and a demographic variance σ_d^2. If $\sigma_e^2 > 0$, then the distribution of abundance is given by

$$\lambda(x) = \frac{\alpha\omega}{1+\varepsilon} \exp\left[-\frac{1}{2}\frac{[\ln(x+\varepsilon) - r/\gamma]^2}{\sigma_e^2/2\gamma}\right],$$

where

$$\varepsilon = \sigma_e^2/\sigma_d^2$$

and

$$\alpha = \frac{2}{\sigma_e^2} \exp\left[\frac{\gamma}{\sigma_e^2}[\ln(1+\varepsilon) - r/\gamma]^2\right].$$

This equation for $\lambda(x)$ is a lognormal model shifted or translated from x to $x + \varepsilon$, where ε is the ratio of environmental to demographic stochastic variance.

Engen and Lande's (1996) model differs in a number of important ways from the neutral model developed in this book. First, it treats the effects of both environmental stochasticity and demographic stochasticity on population dynamics, whereas the theory I develop here thus far only has demographic stochasticity. Second, it incorporates density-dependent population growth without zero-sum dynamics. The theory here assumes zero-sum dynamics that cap maximum population size, but it does not impose per capita restraints on birth and death rates that change with population size. Thus far, their theory does not include migration rates, whereas the present theory does (chapters 4–6). It should be possible to evaluate which theory applies in a particular situation by the relationship between the mean and variance of population sizes across a community. More importantly, although Engen and Lande's theory can produce a lognormal relative abundance distribution with the appropriate choice of model for stochastic density dependence, we have seen that most observed relative abundance distributions are not, in fact, lognormal. It remains to be seen whether their model can produce the asymmetrical zero-sum multinomial distribution of relative species abundance with its observed long tail of very rare species. I now turn my attention to the development of the unified neutral theory of biodiversity and biogeography.

SUMMARY

1. Relatively few models have taken a dynamic approach to a theory of relative species abundance, building on processes of birth, death, migration, and speciation. The first of these dynamical theories, by Caswell, would have performed far better had the assumption of constant community size and zero-sum dynamics been explored.

2. The key first principle of the unified neutral theory is that the dynamics of communities are a zero-sum game. No species can increase in abundance in the community without a matching decrease in the collective abundance of all other species.

3. This principle follows immediately from the generally observed, remarkably precise, linear relationship between the number of individuals and sample area in a community, a relationship that holds quite generally in ecological communities, irrespective of the turnover of species from one local area to another.

4. If the dynamics of species abundances are neutral on a per capita basis under zero sum dynamics, then a theorem can be proven about relative species abundance that predicts that relative abundances will be described by a new statistical distribution called a zero-sum multinomial.

5. The zero-sum multinomial distribution is similar to a lognormal distribution for common species, but it differs in shape for the rare species. Unlike the lognormal, it is asymmetrical, typically with a long tail of very rare species. Most newer, larger datasets on relative species abundance in natural communities exhibit this asymmetry and long tail of rare species.

6. The long tail of very rare species means that relative abundance distributions are not canonical, *sensu* Preston, because there is no fixed relationship between the total number of individuals sampled and the abundance of the rarest species.

7. At least one other dynamical theory of relative species abundance has, with different assumptions, been able to generate lognormal distributions. However, none of the other theories except the present one reproduces the zero-sum multinomial distribution and the long tail of rare species observed in real datasets.

Local Community Dynamics under Ecological Drift

In previous chapters I examined current models of relative species abundance based on niche-assembly theories, and then a small class of dynamical models whose predictions derive from birth, death, and dispersal processes. I now discuss one of the latter models in greater detail, zero-sum ecological drift, the foundation for a unified theory of biodiversity and biogeography. It is useful to divide the problem into two spatio-temporal scales for analysis: local community dynamics, which are relatively rapid, and metacommunity dynamics, which are much slower and occur on large scales. In this chapter, I study local community dynamics from the perspective of a single arbitrary species undergoing ecological drift in the community. In the next chapter, I consider metacommunity dynamics and the coupling of metacommunity and local community dynamics that results in the full multispecies ecological drift process under the unified theory.

Consider a local community saturated with individuals as, for example, trees in a closed-canopy forest. Let the community consist of J trees, regardless of species. Suppose individuals resist displacement until killed by some disturbance. Let each disturbance kill D individuals at random in the community. Let M of these individuals be replaced by immigrants from the metacommunity. Let the $D - M$ local replacement individuals be drawn at random from the species that survive, with probabilities set by their post-disturbance relative abundances.

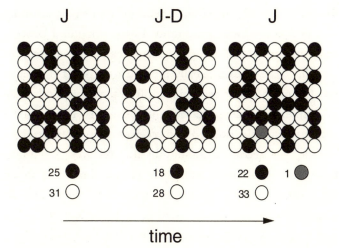

FIG. 4.1. Cartoon of one disturbance cycle in a model community undergoing zero-sum ecological drift. At the beginning of the cycle are two species whose individuals occupy all sites or resources (*left*). Immediately after the disturbance, which killed several individuals of both species, vacant sites or unutilized resources are opened up (*middle*). These are occupied by recruits from the two species in the local community, and by an immigrant individual of a third species from the metacommunity source area (*right*).

The process can be visualized by a simple cartoon (fig. 4.1). The community starts each disturbance cycle completely filled with individuals of the various species in the community (left panel). Then a disturbance occurs, killing *D* individuals, which creates openings or unused resource (middle panel). Then recruitment fills the openings, either with local recruits or immigrants from the metapopulation source area (right panel). A new species recruited a single individual from the source area in this example (gray circle).

The cartoon in figure 4.1 illustrates the full multispecies random walk under zero-sum ecological drift. To present the multispecies theory mathematically, however, it is useful first to analyze the dynamical behavior of a single arbitrary species i in the local community. In this chapter, I concentrate on the fate of the ith species against a backdrop of all other species dynamically lumped together. I then analyze full multispecies ecological drift in the local community in chapter 5. I analyze species i under two scenarios, first a community that is isolated and receives no immigrants from outside, and second a community that receives immigrants from an external metacommunity. Let the number of immigrant individuals per disturbance cycle be M where $M < D$. Thus, the community will undergo independent ecological drift if $M = 0$, and it will be influenced by the relative abundances of species in the source area if $M > 0$. I will use lower case m to denote the per capita or per birth probability that an individual is replaced by an immigrant. Case I is the so-called *absorbing* case, because once species go extinct or reach monodominance there is no further change (so these are final "absorbing" states). Case II is called the *ergodic* case because all states of abundance are reachable from one another, i.e., local extinction and monodominance are not permanent and final states.

Consider the isolated community first, the absorbing case. Let us now follow the fate of a single member species, species i. An important question to answer is: How long does it take an arbitrary species to go extinct or to achieve complete dominance under zero-sum ecological drift? Species drift to the absorbing abundance states of 0 or J. The time to "fixation" or absorption will depend on the size of the community J, the disturbance rate, D, and the initial abundance of the focal species, N_i. I first focus on the analytically more tractable case of $D = 1$. The transition probabilities are

$$\Pr\{N_i - 1|N_i\} = \mu\left(\tfrac{N_i}{J}\right)\left(\tfrac{J-N_i}{J-1}\right)$$

$$\Pr\{N_i|N_i\} = 1 - \mu + \mu\left(\tfrac{N_i}{J}\right)\left(\tfrac{N_i-1}{J-1}\right)$$

$$+ \mu\left(\tfrac{J-N_i}{J}\right)\left(\tfrac{J-N_i-1}{J-1}\right)$$

$$\Pr\{N_i + 1|N_i\} = \mu\left(\tfrac{J-N_i}{J}\right)\left(\tfrac{N_i}{J-1}\right),$$

where μ is the probability of one death per time step. A numerically equivalent and simpler combinatorial form for the transition probabilities is

$$\Pr\{N_i - 1|N_i\} = \Pr\{N_i + 1|N_i\} = \frac{\binom{J-2}{N_i-1}}{\binom{J}{N_i}}\mu$$

$$\Pr\{N_i|N_i\} = \frac{\binom{J}{N_i} - 2\mu\binom{J-2}{N_i-1}}{\binom{J}{N_i}}.$$

For simplicity and with no loss of generality, I will hereafter rescale time by the average death rate in the community, so that one death occurs per time step (i.e., $\mu = 1$). The Markovian matrix M for the absorbing case of zero-sum ecological drift for focal species i and for the special case for $D = 1$, is

$$M = \begin{pmatrix}
1 & 0 & 0 & 0 & \cdots & 0 & 0 & 0 & \cdots 0 & 0 & 0 \\
\frac{1}{J} & \frac{J-2}{J} & \frac{1}{J} & 0 & \cdots & 0 & 0 & 0 & \cdots 0 & 0 & 0 \\
0 & \frac{2(J-2)}{J(J-1)} & \frac{J(J-1)-4(J-2)}{J(J-1)} & \frac{2(J-2)}{J(J-1)} & \cdots & 0 & 0 & 0 & \cdots 0 & 0 & 0 \\
\vdots & \vdots & \vdots & \vdots & \vdots & \vdots & \vdots & \vdots & \vdots \vdots & \vdots & \vdots \\
0 & 0 & 0 & 0 & \cdots & \frac{\binom{J-2}{k-1}}{\binom{J}{k}} & \frac{\binom{J}{k}-2\binom{J-2}{k-1}}{\binom{J}{k}} & \frac{\binom{J-2}{k-1}}{\binom{J}{k}} & \cdots 0 & 0 & 0 \\
\vdots & \vdots & \vdots & \vdots & \cdots & \vdots & \vdots & \vdots & \cdots \vdots & \vdots & \vdots \\
0 & 0 & 0 & 0 & \cdots & 0 & 0 & 0 & \cdots \frac{1}{J} & \frac{J-2}{J} & \frac{1}{J} \\
0 & 0 & 0 & 0 & \cdots & 0 & 0 & 0 & \cdots 0 & 0 & 1
\end{pmatrix}$$

This transition probability matrix is square with $J + 1$ columns and rows corresponding to the k possible abundance states, 0 to J, of the ith species. The rows correspond to the abundance of the ith species at time t and the columns to its abundance at time $t + 1$. So, for example, the entry in row 2, column 1, corresponds to the probability that the ith species, starting at abundance 1 at time t will go extinct (have abundance 0) at time $t + 1$, which is $1/J$. Most of the entries in the matrix are zero because we allow only one death and one birth per time step. Thus, there are probabilities along the principal diagonal, which correspond to no change in abundance, and immediately on either side of the principal diagonal, corresponding to a decrease or an increase in abundance of one individual in one time step. Note that the probabilities across the columns in a given row sum to unity. If $N(t)$ is a row vector of probabilities that species i is at abundances 0 through J at time t, then the row vector of probabilities at time $t + 1$ can be found simply as

$$N(t + 1) = N(t) \cdot M.$$

For a very clear introduction to the matrix methods used hereafter in this chapter, and the fundamental theorems needed for analyzing the behavior of Markovian stochastic processes, I recommend the excellent book, *Finite Markov Chains*, by Kemeny and Snell (1960).

Matrix M can be put into the general canonical form of an absorbing Markov chain by arranging the two absorbing states of abundances 0 and J to be the first two states (upper left corner). Then matrix M can be partitioned into four submatrices,

$$M = \begin{pmatrix} I & 0 \\ R & Q \end{pmatrix},$$

where I is the identity matrix representing the submatrix of absorbing states. Submatrix Q represents transitions among transient abundances, 1 through $J - 1$. Over Q is a submatrix of zeros (because by definition the absorbing abundance states of 0 and J cannot be left), and submatrix R represents the final transitions to the absorbing states. The rearranged matrix M is the standard form of an absorbing Markov chain. Then the fundamental matrix A of the absorbing ecological drift process for focal species i is given by

$$A = (I - Q)^{-1}.$$

The elements $a_{j,k}$ of fundamental matrix A give the expected number of times that species i passes through abundance N_k having started in abundance N_j, before extinction or complete dominance. Matrix A allows us to determine the total number of births and deaths that will occur in the community before extinction or complete dominance, starting with abundance N_i. The vector $T(N)$ of fixation times is

$$T(N) = A\zeta,$$

where ζ is a column vector with all entries equal to unity. After some algebraic manipulation, one can show for the special, sparse-matrix case of $D = 1$ that

$$T(N_i) = (J - 1)\left[(J - N_i) \sum_{k=1}^{N_i} (J - k)^{-1} + N_i \sum_{k=N_i+1}^{J-1} k^{-1} \right]$$

(Hubbell and Foster 1986a). The first term in this expression is the mean time that species i is expected to spend at abundances less than or equal to N_i, and the second term is the mean time species i will spend at abundances greater than N_i before absorption. Note that these times are relative and are measured in the total number of deaths happening in all species in the community. The vector of variances $\text{Var}(T)$ in the time to absorption is also computed from the fundamental matrix A:

$$\text{Var}(T) = (2A - I)T - T_{\text{sq}},$$

81

where T_{sq} is the vector $T(N)$ with every element squared. Unfortunately, Var(T) is not a compact expression to write down as an explicit algebraic function of N_i and J, but it is numerically straightforward to compute its exact value.

I have illustrated how the time to extinction or complete dominance varies as a function of community size J and initial population size N_i in figure 4.2. Here the time to extinction or monodominance is measured in terms of the total number of deaths that occur in all species in the community prior to fixation of the ith species. The curves illustrate how this time varies in community sizes ranging from $J = 4$ to $J = 128$ individuals. The time to extinction or complete dominance is maximal when the initial abundance of the focal species that is $J/2$, and the curves are symmetrical about this abundance, with their shortest times when $N_i = 1$ and $N_i = J - 1$.

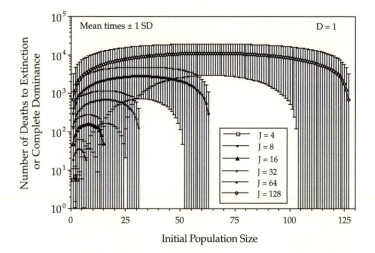

FIG. 4.2. Number of all deaths until fixation (extinction or monodominance) of arbitrary species i as a function of community size J. Note that the figure is a semilogarithmic plot. The longest times to fixation are from an initial species abundance of $J/2$.

The most important conclusion we can draw from figure 4.2 is that the time to extinction or complete dominance rises rapidly with community size and size of the focal population (Hubbell and Foster 1986a). If the ith species represents a small fraction of the total community size, i.e., $N_i \ll J$, then $T(N_i) \cong N_i(J - 1)[1 + \ln(J)]$. Thus, under the special case of $D = 1$, the time to extinction or monodominance grows as the product of the population size of the given species, times the community size, times the log of community size! This is potentially a very large number. For example, if a species comprises 10% of a community having 2000 individuals of all species, then the community will completely turn over almost seven hundred times before the species is expected to go extinct or become completely dominant. This represents nearly 1.3 million individuals that die and are replaced.

Thus, pure ecological drift to extinction can be a slow process if only a single individual is killed per disturbance. However, as we shall see later in this chapter, when $D \gg 1$, species can go locally extinct in much shorter amounts of time. The minimum time to extinction for the ith species is clearly N_i/D time steps, where N_i is the initial abundance of the species. For this reason, frequencies of times to extinction or complete dominance do not simply decline monotonically with a mode at $t = 0$, but instead are approximately gamma distributed (fig. 4.3). There is an abrupt rise to the modal value, followed by a much slower, nearly exponential decay in the frequency of fixation with increasing time. Note, however, that community dynamics under ecological drift are considerably faster if the disturbance rate is much higher ($D \gg 1$), as will be discussed later. Very short times to extinction are also likely to be prevalent when there is high demographic or environmental variance in the death rate.

I now turn to the second case, namely, a community that is open to immigration ($M > 0$). In this case the metacommunity may serve as a source of immigrants for

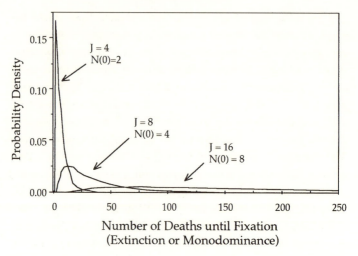

FIG. 4.3. Probability density functions for the time to fixation (extinction or complete dominance), illustrated as a function of community size J. The distributions are approximately gamma. The cases illustrated are for the longest times to fixation for the given community sizes, namely from an initial abundance of $J/2$.

any given species into the local community. The population dynamics of each species are once again Markovian, but now they are ergodic because extinction and monodominance in the local community are no longer absorbing states. Thus, under ecological drift, monodominant communities can be invaded by other species. Species are expected to go locally extinct and reimmigrate repeatedly from the source area. Unlike the absorbing case without immigration, species will have a positive expected equilibrium abundance in the ergodic community, provided that the given species also has nonzero abundance in the metacommunity.

The most fundamental question that one seeks to answer initially is: What is the expected abundance of a given species in the community at stochastic equilibrium, and what is its variance? A second related question is: What is the incidence of a given species in the ergodic community under zero-sum ecological drift? To answer these questions, we need to compute the abundance eigenvector of the limiting proportion of time spent at each abundance from 0 to J as $t \to \infty$. The eigenvector is very important because we can calculate directly from it the theoretically expected abundance of any species and its variance, as a function of community size, immigration rate, disturbance rate, and the relative abundance of the species in the metacommunity. The column eigenvector ψ is found by solving the equation $\psi^T B = \psi^T$ subject to the constraint that $\sum_{n=0}^{J} \psi(n) = 1$, where B is the transition matrix of the ergodic community. Note that we are solving for the *left* eigenvector of the transition matrix because by convention we have chosen the rows to be the current abundance state (at time t) and the columns to be the next abundance state at time $t + 1$. This is not the usual convention, but it is used by Kemeny and Snell (1960), and I like it because it generates a more natural flow computationally from left to right. The expected local abundance of the ith species and its variance are simply

$$E\{N_i\} = \sum_{n=1}^{J} n \cdot \psi(n)$$

$$\mathrm{Var}\{N_i\} = \sum_{n=1}^{J} (n - E\{N_i\})^2 \cdot \psi(n).$$

The incidence of a species is the proportion of time that the species is present in the local community. This proportion is $\sum_{n=1}^{J} \psi(n) = 1 - \psi(0)$, the sum of the eigenvector elements comprising all nonzero abundances. Given the important quantities that one can compute from the abundance eigenvector, it is well worth the effort to derive its algebraic expression for an ergodic community of arbitrary size J. However, I will defer further discussion of incidence functions until chapter 7. Once again assuming that exactly one death and one birth occur per time step, the transition probabilities for the ith species abundance N_i in the ergodic community with immigration are

$$\Pr\{N_i - 1 | N_i\} = \frac{N_i}{J}\left[m(1 - P_i) + (1 - m)\left(\frac{J - N_i}{J - 1}\right)\right]$$

$$\Pr\{N_i | N_i\} = \frac{N_i}{J}\left[m P_i + (1 - m)\left(\frac{N_i - 1}{J - 1}\right)\right] + \left(\frac{J - N_i}{J}\right)$$

$$\times \left[m(1 - P_i) + (1 - m)\left(\frac{J - N_i - 1}{J - 1}\right)\right]$$

$$\Pr\{N_i + 1 | N_i\} = \left(\frac{J - N_i}{J}\right)\left[m P_i + (1 - m)\left(\frac{N_i}{J - 1}\right)\right],$$

where m is the probability that a death will be replaced by an immigrant, and P_i is the fractional metacommunity relative species abundance of the ith species. It is easy to walk through these equations in words. For example, the first equation gives the transition probability for the ith species to decline in abundance by one individual. For this to happen, a death must occur in the ith species, N_i / J, and the birth must be in some other species. The first probability inside the brackets is that of an immigration event of some species other than i: $m(1 - P_i)$. The second probability is that of having no immigration event and a local birth in a species other than i: $(1 - m)(J - N_i)/(J - 1)$. Note that when the immigration rate, m, is zero, these probabilities reduce to the absorbing case.

The slow dynamics of the much larger metacommunity source area allows us temporarily to treat the distribution of P as a fixed marginal distribution. The full theory, which includes metacommunity dynamics, does not require specification of any marginal distributions (chapter 5). In the full unified theory, relative species abundances in the local community and in the metacommunity are always predictions, never givens. Eliminating the marginal distribution will obviate the need for any species-specific parameters in the unified theory.

The simplest possible ergodic community is a community of size 1, representing the dynamics of replacement of a single individual. In this case the equations of the ergodic community reduce to the following:

$$B = \begin{pmatrix} 1 - mP_i & mP_i \\ m(1 - P_i) & 1 - m(1 - P_i) \end{pmatrix},$$

which has eigenvector $\psi_i(n) = \Pr\{0, 1\} = \{1 - P_i, P_i\}$. The entries in this simple transition matrix are easy to obtain from the transition probability equations. For example, the entry in row 1, column 1, is the probability of being at zero abundance at time t and remaining at zero abundance at time $t + 1$. From the second element of the eigenvector, the expected fraction of time that the ith species will be present at equilibrium is simply its metacommunity relative species abundance. Thus, at the scale of replacing a single individual, the presence or absence of a species must clearly be independent of the immigration rate, m. For larger communities, the abundance eigenvector is no longer independent of the probability of immigration, m. For example, for $J = 3$, the eigenvector is

$$\psi_i(n) = \Pr \begin{Bmatrix} N_i = 0 \\ N_i = 1 \\ N_i = 2 \\ N_i = 3 \end{Bmatrix}$$

$$= \left(\frac{1}{1+m}\right) \begin{pmatrix} (1 - P_i)(1 - mP_i)(1 + m - 2mP_i) \\ 3mP_i(1 - P_i)(1 + m - 2mP_i) \\ 3mP_i(1 - P_i)(1 - m + 2mP_i) \\ P_i[1 - m(1 - P_i)](1 - m + 2mP_i) \end{pmatrix}.$$

The reason for writing the eigenvector in this partially unfactored form will be clear when the eigenvector for the community of arbitrary size J is found.

CHAPTER FOUR

For $J = 3$, the expected abundance of the ith species is

$$E\{N_i|J = 3\} = \left(\frac{1}{1+m}\right)\{0 \cdot (1 - P_i)(1 - mP_i)(1 + m - 2mP_i)$$
$$+ 1 \cdot 3mP_i(1 - P_i)(1 + m - 2mP_i)$$
$$+ 2 \cdot 3mP_i(1 - P_i)(1 - m + 2mP_i)$$
$$+ 3 \cdot P_i[1 - m(1 - P_i)](1 - m + 2mP_i)\},$$

which simplifies to $3P_i$. The variance is

$$\text{Var}\{N_i|J = 3\} = \left(\frac{1}{1+m}\right)\{(0 - 3P_i)^2(1 - P_i)(1 - mP_i)$$
$$\times (1 + m - 2mP_i) + (1 - 3P_i)^2 3mP_i(1 - P_i)$$
$$\times (1 + m - 2mP_i) + (2 - 3P_i)^2 3mP_i(1 - P_i)$$
$$\times (1 - m + 2mP_i) + (3 - 3P_i)^2 P_i$$
$$\times [1 - m(1 - P_i)](1 - m + 2mP_i)\},$$

which simplifies to $3P_i(1 - P_i)(3 - m)/(1 + m)$.

If $P_i = 0$, the ith species is extinct in the metacommunity, and its expected abundance in the local community is zero with zero variance. Therefore, metacommunity extinction implies local community extinction. When $P_i = 1$, the species is monodominant in both the metacommunity and in the local community, with zero variance. For variation in P_i, the variance in abundance is maximal when $P_i = (9 - 3m)/6(3 - m) = 1/2$. For variation in immigration m, the variance is maximized as $m \to 0$. Thus, species abundances are more variable in local communities that are more isolated from the metacommunity.

Now consider ergodic communities of arbitrary size J. After six weeks of algebra, we find that the general solution

88

eigenvector is

$$\psi_i(n) = \Pr\left\{\begin{array}{c} 0 \\ 1 \\ \vdots \\ k \\ \vdots \\ J-1 \\ J \end{array}\right\} =$$

$$\left(\begin{array}{c} \dfrac{\binom{J}{0}(1-P_i)(1-mP_i)\prod_{x=1}^{J-2}G(J,m,P_i,x)}{\prod_{x=1}^{J-2}[(J-1)-x(1-m)]} \\[3ex] \dfrac{\binom{J}{1}(1-P_i)mP_i\prod_{x=1}^{J-2}G(J,m,P_i,x)}{\prod_{x=1}^{J-2}[(J-1)-x(1-m)]} \\[3ex] \vdots \\[1ex] \dfrac{\binom{J}{k}(1-P_i)mP_i\prod_{x=1}^{k-1}H(J,m,P_i,x)\prod_{x=k}^{J-2}G(J,m,P_i,x)}{\prod_{x=1}^{J-2}[(J-1)-x(1-m)]} \\[3ex] \vdots \\[1ex] \dfrac{\binom{J}{J-1}(1-P_i)mP_i\prod_{x=1}^{J-2}H(J,m,P_i,x)}{\prod_{x=1}^{J-2}[(J-1)-x(1-m)]} \\[3ex] \dfrac{\binom{J}{J}P_i[1-m(1-P_i)]\prod_{x=1}^{J-2}H(J,m,P_i,x)}{\prod_{x=1}^{J-2}[(J-1)-x(1-m)]} \end{array}\right),$$

where

$$G(j,m,P_i,x) = (J-1)(1-mP_i) - x(1-m)$$

and

$$H(j,m,P_i,x) = (J-1)mP_i + x(1-m).$$

The eigenvector gives the probability density for any abundance $0 \le N_i \le J$, for the general ergodic community of size J. From the eigenvector we can calculate the equilibrium abundance of the ith species in the community and its variance. We find that the expected abundance of the ith species is

$$E\{N_i\} = \sum_{k=0}^{J} \psi(k) \cdot k = JP_i.$$

Thus, the expected abundance of the ith species at equilibrium in the local community is simply equal to the local community size, J, times the metacommunity relative abundance of the ith species, P_i. The mean local abundance of the ith species is thus independent of immigration rate, m. However, as we will see shortly, if m is small, then the ith species will spend a very low proportion of time at its mean abundance.

The variance depends on all three parameters, J, m, and P_i, and is given by

$$\text{Var}\{N_i\} = \sum_{k=1}^{J} (k - E\{N_i\})^2 \cdot \psi(k) =$$

$$\frac{\sum_{k=0}^{J} \left[C(J, m, P_i, k) \cdot \prod_{x=1}^{k-1} H(J, m, P_i, k) \cdot \prod_{x=k}^{J-2} G(J, m, P_i, k) \right]}{\prod_{x=1}^{J-2} (J-1) - x(1-m)},$$

where

$$C(J, m, P_i, k) =$$
$$\begin{cases} \binom{J}{k}(JP_i)^2(1-P_i)(1-mP_i) & \text{for } k = 0 \\ \binom{J}{k}(k-JP_i)^2(1-P_i)mP_i & \text{for } k = 1, 2, \ldots, J-1 \\ \binom{J}{k}(k-JP_i)^2 P_i[1-m(1-P_i)] & \text{for } k = J \end{cases}$$

and

$$\prod_{x>k-1}^{k-1} H(J, m, P_i, x) = 1 \quad \text{and} \quad \prod_{x>J-2}^{J-2} G(J, m, P_i, x) = 1.$$

We can now explore the behavior of the abundance eigenvector and the mean and variance of species i. The probability density functions for the abundance of species i in the local community reveal how local abundance depends on metacommunity abundance P_i, as well as on immigration rate, m, and community size, J. Figure 4.4 illustrates a family of probability density functions varying metacommunity abundance, for a community of size $J = 64$ and immigration rate of $m = 0.05$. Superficially these lines look like dominance-diversity curves, but they describe the probability that an arbitrary species i will be at a given abundance on the x-axis. For example, when P_i is large (e.g., 0.999),

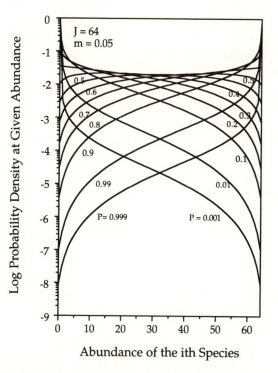

FIG. 4.4. The effect of varying the metacommunity relative abundance of the ith species on its equilibrium probability density functions in an ergodic local community undergoing zero-sum ecological drift. Numerical example of a community of $J = 64$ individuals and an immigration probability of $m = 0.05$ per replacement.

such that the species is almost monodominant in the meta-community, it spends most of its time monodominant in the local community as well. Conversely, when P_i is small (e.g., 0.001), such that it is rare in the metacommunity, then it will almost always be absent from the local community. In all distributions, the expected abundance of the ith species in the local community is JP_i. However, when P_i is large, the probability is concentrated at high abundance, and conversely when P_i is small, the probability is concentrated at low abundance. Note the complementary shape to the probability density functions at high and low P_i. The family of curves in figure 4.4 happens to be fairly symmetrical for this particular case of $m = 0.05$, but the curves become less symmetrical as the immigration rate m is varied up or down.

The probability density function exhibits richer behavior when the immigration rate is varied (fig. 4.5). When the local community is very isolated from the metacommunity, such that immigration rate m is small, then the abundance probability density functions are U-shaped. This shape arises because ecological drift has ample time in between rare immigration events to carry the relative abundance of the ith species to 0 (locally extinct) or to 1 (monodominance) when the community is quite isolated. As the immigration rate increases, local community dynamics are more strongly coupled to the metacommunity. As the immigration rate increases, the probability density function becomes unimodal, and the limiting variance in abundance is reduced. As $m \to 1$, the mode of the density function is near JP_i, which, is equal to 6.4 ($P_i = 0.1$) in the case illustrated in figure 4.5.

The effect of increasing community size J while holding m and P_i constant is shown in figure 4.6. As J increases, the modal abundance of the probability density function increases, but the distribution also broadens considerably, and the probability of the most frequent abundance becomes lower. This is expected simply because there

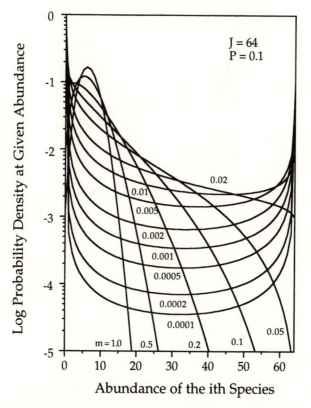

FIG. 4.5. The effect of varying the probability of immigration m on the equilibrium probability density functions for the ith species in an ergodic community undergoing zero-sum ecological drift. $J = 64$ individuals and $P_i = 0.1$.

are more abundance states that can be occupied in larger communities.

It is useful to illustrate the dynamical behavior of the ith species under ecological drift, to demonstrate the reality of the stochastic equilibria obtained analytically from the abundance eigenvector. Consider a numerical example of a very small local community consisting of just 16 individuals. Figure 4.7, top panel, shows the random walk of a species

93

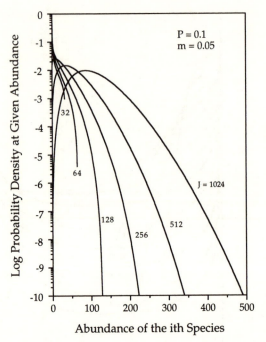

FIG. 4.6.   The effect of varying community size, J, on the equilibrium probability density functions for the ith species in an ergodic community undergoing zero-sum ecological drift. P_i and m are held constant at 0.1 and 0.5, respectively. Note that the mode of the density function approximately tracks $J \cdot P_i$.

that has relative abundances P_i in the metacommunity of 0.95 or of 0.05, respectively. The bottom panel tracks the abundance of a species with a metacommunity relative abundance of 0.5. Note the greater variance for $P_i = 0.5$.

The variability of abundance of a species about the stochastic equilibrium JP_i is less in larger communities that are not disturbed by high death rates. Figure 4.8 shows a species drifting in abundance in a local community of size $J = 64$, experiencing low, medium, and high disturbance, and which is increasingly isolated from the source area. In each case the metacommunity relative abundance

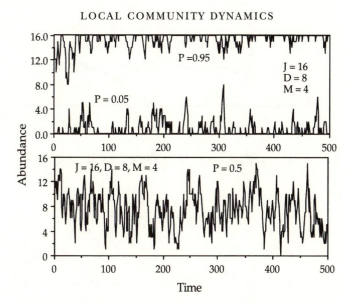

FIG. 4.7. Random walk of an arbitrary species in a small community ($J = 16$). *Top*: $P_i = 0.95$ and 0.05. *Bottom*: $P_i = 0.5$. The dynamical behavior appears especially erratic in small communities. In these examples, the death rate per disturbance cycle was set high ($D = 8$). *Note*: The capitals D and M in these graphs are the actual number of individuals dying and immigrating per disturbance cycle, respectively. The probability of immigration m is equal to M/D.

of the focal species is 0.5. When a single individual is killed and replaced from the metacommunity (no isolation) during each disturbance cycle (top panel), the abundance of the species fluctuates around the equilibrium value of 32 by about ± 5 individuals over 500 birth-death cycles. However, when half the individuals in the local community are killed and replaced during each cycle, but only one in 32 replacement individuals is an immigrant from the source area, then the relative abundances fluctuate wildly. In this run of the ecological drift model, the species even spends some time as a monodominant and hardly shows any central tendency about its expected abundance of 32 individuals in 500 disturbance cycles.

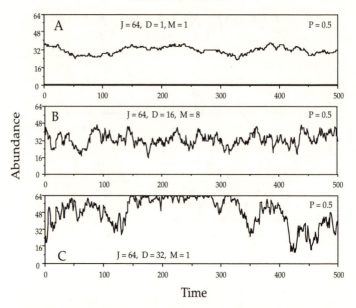

FIG. 4.8. Effect of varying the death rate, D, and the probability of immigration, $m = M/D$, where $M \leq D$. Community size, $J = 64$. *Top*, low disturbance: $D = 1 M = 1$. *Middle*, medium disturbance: $D = 16$, $M = 8$. *Bottom*, high disturbance: $D = 32$, $M = 1$. Once again, D and M in these graphs are the actual number of individuals dying and immigrating per disturbance cycle, respectively.

The theory of ecological drift predicts that the variance in local abundance of the ith species will be a parabolic function of metacommunity abundance P_i, with maximal variance when $P_i = 0.5$. The curve is parabolic when varying P_i, because a common factor in the variance is a quadratic function of P_i, namely $JP_i(1 - P_i)$. The variance as a function of P_i is shown in figure 4.9 for various values of the immigration rate m. The variance in the local abundance of the ith species increases dramatically for small m, i.e., for very isolated islands or local communities, a result that would be expected from figure 4.5. Figure 4.9 is a semilogarithmic plot of the variance and mean; these are inverted parabolas, log transformed, for various values of m.

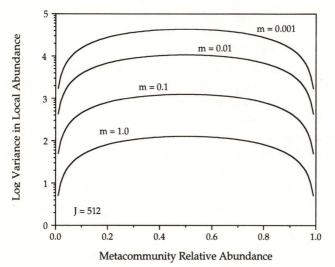

FIG. 4.9. Predicted relationship between the local variance in abundance of the ith species and its metacommunity relative abundance, for various values of the immigration rate, m. The curves are inverted parabolas that are distorted by the log transformation of the y-axis.

The quadratic curvilinearity of this relationship is not predicted by stochastic models that do not assume a zero-sum rule. Note that a quadratic relationship is also expected between the mean and variance in local abundance because the expected local abundance is proportional to P_i.

More generally, if there is a significant quadratic signature in the relationship of mean species relative abundances and their variances in a time series for a local community, this would constitute evidence in support of zero-sum dynamics. In practice, detecting this quadratic signature may be very difficult. This is because full parabolic variance curves will rarely be observed in nature. In most natural metacommunities, even the most abundant species will rarely constitute more than a few percent of the total metacommunity. If the commonest metacommunity species represents, say, less than 10% of the metacommunity, then the theory

97

of ecological drift predicts that there will be nearly a perfect log-log linear relationship between mean and variance in abundance of the ith species. Figure 4.10 is a plot of log variance versus log mean for $P_i < 0.10$ and various values of m. Note that under log transformation, the slopes become essentially independent of m.

Relatively long-term data are available on the temporal dynamics of a community of more than 300 species of Macrolepidoptera in southern England. These data can be used to illustrate the relationship between variance and mean abundance predicted by the theory. Permission to use the data for these analyses was kindly given by Laurence

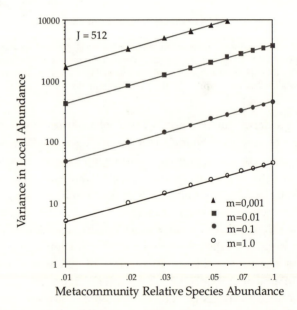

FIG. 4.10. Relationship between the log of the local variance in abundance of the ith species versus metacommunity abundance for species whose metacommunity abundances represent less than 10% of total metacommunity size, showing the log-log linearity. The intercepts but not the slopes are functions of the immigration rate m on the log-transformed plot.

Cook of the Manchester Museum. For 25 years, moths were collected at mercury vapor light traps in Woodchester Park Field Centre near Manchester, beginning in the late 1960s (Baker 1985, Cook and Graham 1996).

The relationship between log variance and log mean in percentage of sample abundance of 306 species is shown in figure 4.11, for all species that occurred in at least 7 years out of 25. Species with lower incidence were excluded because the variance is largely driven by presence-absence for very infrequent species. There is no significant quadratic term in the relationship. If these data do reflect zero-sum ecological drift, then the "true" metacommunity being sampled must be truly enormous with many species, each representing a small fraction of the metacommunity.

Thus far I have considered only the mildest form of disturbance to the community—the death and replacement of

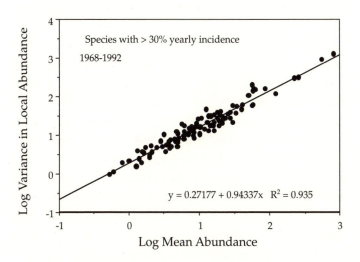

FIG. 4.11. Relationship between the log of mean annual abundance and variance in annual abundance for Macrolepidoptera collected over a 25-year period in Gloucestershire, England. Original data provided by L. C. Cook.

a single individual at a time ($D = 1$). I now turn to the analytically more difficult cases of $D > 1$. As I will now show, persistence times—the time it takes a drifting species to go extinct in the local community—become shorter, and often much shorter, when the community is subject to more severe disturbances. The matrices for ergodic communities having $D > 1$ are also analytically solvable in the numerical sense, but their eigenvectors are much more complex and difficult to find algebraically. Nevertheless, in what follows the results presented are all exact numerical solutions to the equations. In spite of their difficulty, cases of $D > 1$ are important to study because natural communities are subject to different average rates of mortality as well as environmental stochasticity in the mortality rates themselves. Moderate to large environmental disturbances strike virtually every community at least occasionally. Communities can be suddenly and massively disturbed by events such as intense storms, landslides, fires, earthquakes, and epidemics (Lande 1993).

In this chapter I consider only demographic stochasticity because D is treated as a fixed parameter. However, as it turns out, the mean abundance of the ith species is identical under environmental variance in the death rate D, so long as zero-sum dynamics apply and the average death rate remains D. Indeed, the expected distribution of relative species abundance in the local community also remains unchanged under stochastically varying D, although the variance is affected. Thus, the analytical results for expected abundances derived under a constant death rate apply equally well to zero-sum communities subject to stochastically variable death rates having the same mean.

Consider first the case of $D = 2$. This case permits the ith species to change ± 2 individuals per disturbance cycle, provided and $N_i \geq 2$ and $\leq J - 2$. There are now five potentially nonzero transition probabilities:

$$\Pr\{N_i - 2 | N_i\} = \frac{N_i(N_i - 1)}{J(J - 1)} \left[\begin{array}{c} m^2(1 - P_i)^2 + 2m(1 - m) \\ \times (1 - P_i)\left(\frac{J - N_i}{J - 2}\right) \\ + (1 - m)^2\left(\frac{J - N_i}{J - 2}\right)^2 \end{array} \right]$$

$$\Pr\{N_i - 1 | N_i\} = \frac{N_i(N_i - 1)}{J(J - 1)} \left[\begin{array}{c} 2m^2 P_i(1 - P_i) \\ + 2m(1 - m)P_i\left(\frac{J - N_i}{J - 2}\right) \\ + 2m(1 - m)(1 - P_i) \\ \times \left(\frac{N_i - 2}{J - 2}\right) + 2(1 - m)^2 \\ \times \left(\frac{N_i - 2}{J - 2}\right)\left(\frac{J - N_i}{J - 2}\right) \end{array} \right]$$

$$+ 2\frac{N_i(J - N_i)}{J(J - 1)} \left[\begin{array}{c} m^2(1 - P_i)^2 \\ + 2m(1 - m) \\ \times (1 - P_i)\left(\frac{J - N_i - 1}{J - 2}\right) \\ + (1 - m)^2\left(\frac{J - N_i - 1}{J - 2}\right)^2 \end{array} \right]$$

$$\Pr\{N_i | N_i\} = \frac{N_i(N_i - 1)}{J(J - 1)} \left[\begin{array}{c} m^2 P_i^2 + 2m(1 - m)P_i\left(\frac{N_i - 2}{J - 2}\right) \\ + (1 - m)^2\left(\frac{N_i - 2}{J - 2}\right)^2 \end{array} \right]$$

$$+ 2\frac{N_i(J - N_i)}{J(J - 1)} \left[\begin{array}{c} 2m^2 P_i(1 - P_i) + 2m(1 - m) \\ \times P_i\left(\frac{J - N_i - 1}{J - 2}\right) + 2m(1 - m) \\ \times (1 - P_i)\left(\frac{N_i - 1}{J - 2}\right) \\ + 2(1 - m)^2\left(\frac{N_i - 1}{J - 2}\right)\left(\frac{J - N_i - 1}{J - 2}\right) \end{array} \right]$$

$$+ \frac{(J - N_i)(J - N_i - 1)}{J(J - 1)}$$

$$\times \left[\begin{array}{c} m^2(1 - P_i)^2 + 2m(1 - m)(1 - P_i) \\ \times \left(\frac{J - N_i - 2}{J - 2}\right) + (1 - m)^2\left(\frac{J - N_i - 2}{J - 2}\right)^2 \end{array} \right]$$

$$\Pr\{N_i + 1|N_i\} = \frac{(J - N_i)(J - N_i - 1)}{J(J - 1)}$$

$$\times \left[\begin{array}{l} 2m^2 P_i(1 - P_i) + 2m(1 - m)P_i \\ \quad \times \left(\frac{J - N_i - 2}{J - 2}\right) + 2m(1 - m)(1 - P_i) \\ \quad \times \left(\frac{N_i}{J - 2}\right) + 2(1 - m)^2\left(\frac{N_i}{J - 2}\right)\left(\frac{J - N_i - 2}{J - 2}\right) \end{array} \right]$$

$$+ 2\frac{N_i(J - N_i)}{J(J - 1)}$$

$$\times \left[\begin{array}{l} m^2 P_i^2 + 2m(1 - m)P_i\left(\frac{N_i - 1}{J - 2}\right) \\ + (1 - m)^2\left(\frac{N_i - 1}{J - 2}\right)^2 \end{array} \right]$$

$$\Pr\{N_i + 2|N_i\} = \frac{(J - N_i)(J - N_i - 1)}{J(J - 1)}$$

$$\times \left[\begin{array}{l} m^2 P_i^2 + 2m(1 - m)P_i\left(\frac{N_i}{J - 2}\right) \\ + (1 - m)^2\left(\frac{N_i}{J - 2}\right) \end{array} \right].$$

Although tedious, the verbal interpretation of these transition probabilities is straightforward. For example, the first equation is the probability that the ith species will suffer a loss of two individuals in the next disturbance cycle. For this to happen with $D = 2$, both deaths must occur in the ith species (note sampling without replacement) and the births or immigrants must be of other species. The probability of two deaths in species i is then multiplied by the sum of three probabilities, each representing a unique way for the deaths to be replaced by other species. The first expression inside the brackets is the probability of two immigrant individuals of species other than i; the second expression is the probability of one immigrant and one local birth, neither of which is species i (times two combinations); and the third expression is the probability of two local births occurring in species other than i.

Continuing in this manner, we can fully generalize the process to accommodate disturbances of arbitrary size, $1 \leq D \leq J$. Let us now calculate an arbitrary transition probability in which the ith species suffers a net loss of d individuals, $d \leq D$. Let us divide the problem into two parts. We will first calculate the death process and secondly the replacement process. If x deaths befall species i, then $D - x$ deaths must occur in all other species. The probability $\mu(x)$ that species i will suffer the death of x individuals is hypergeometrically distributed:

$$\mu(x) = \Pr\{x | N_i\} = \frac{\binom{N_i}{x}\binom{J - N_i}{D - x}}{\binom{J}{D}}$$

for $x \leq N_i$, and $\mu(x) = 0$ for $x > N_i$. If species i is to experience a net loss of d individuals in the current disturbance cycle, then the *minimum* number of deaths that the ith species can suffer is also d. We therefore must consider all terms of $\mu(x)$ in which $d \leq x \leq D$. If $x = d$, then all replacement individuals must be of species other than i (since species i must show a net loss of d individuals). However, if $x > d$, then precisely $x - d$ of the replacement individuals must be of species i, and the remainder $D - x + d$ must be of species other than i, in order for the ith species to show a net loss of exactly d individuals.

The replacement process consists of two subprocesses: immigration from the source metacommunity, or a local birth, either of which can yield a new individual of the ith species or of some other species. We now calculate the probability of each possible combination of immigrants or local births that comprise $x - d$ replacements of species i and $D - x + d$ replacements of species other than i. Since each of these possible ways of replacing the D total deaths is mutually exclusive and unique, these probabilities are summed. The sum of replacement probabilities is then multiplied by the hypergeometric death probability $\mu(x)$ for the given x value. Finally, these products are summed from

103

$x = d$ to $x = D$, yielding the total probability that species i will suffer a net loss of d individuals.

To formalize this verbal recipe for the replacement process, let us make some variable substitutions to simplify notation. There are four essential probabilities in the replacement process. Let ϕ_i be the probability of an individual of species i immigrating; let $\phi_{\bar{\imath}}$ (with subscript"not i", $\bar{\imath}$) be the probability of some other species immigrating; let λ_i be the probability of a local birth in species i, and let $\lambda_{\bar{\imath}}$ be the probability of a local birth in some other species. The immigration probabilities are independent of the distribution of local deaths between species i and other species, and depend only on the probability of immigration and metacommunity relative species abundances: $\phi_i = mP_i$ and $\phi_{\bar{\imath}} = m(1 - P_i)$, respectively. However, the local birth probabilities are functions of how many deaths have just occurred in the disturbance cycle in the local population of species i and in the collective local population of all other species. If x local deaths occurred in species i, then $D - x$ local deaths occurred in other species. Therefore, the local birth probabilities are

$$\lambda_i(x) = (1 - m)\left(\frac{N_i - x}{J - D}\right)$$
$$\lambda_{\bar{\imath}}(D - x) = (1 - m)\left(\frac{J - N_i - D + x}{J - D}\right).$$

Recall that we are currently calculating the probability that species i will decline in abundance by d individuals in the next disturbance cycle. In this case, as noted above, precisely $x - d$ of the replacement individuals must be of species i, and the remainder $D - x + d$ must be of species other than i. These replacement individuals can be distributed in any combination of immigrants and local births. Therefore, we can now write down the probability that species i will suffer a loss of d individuals, $d \leq D$, in the next time step, as follows:

$$\Pr\{N_i - d|N_i\} =$$

$$\sum_{x=d}^{D} \mu(x) \cdot \left[\sum_{y=0}^{x-d} \sum_{z=y+1}^{D-x+d} \frac{D!}{y!(x-d-y)!z!(D-x+d-z)!} \phi_i^y \right.$$
$$\left. \cdot [\lambda(x)]_i^{x-d-y} \cdot \phi_{\bar{i}}^z \cdot [\lambda(D-x)]_{\bar{i}}^{D-x+d-z} \right].$$

An analogous train of logic yields the transition probability for $\Pr\{N_i + d|N_i\}$:

$$\Pr\{N_i + d|N_i\} =$$

$$\sum_{x=d}^{D-d} \mu(x) \cdot \left[\sum_{y=0}^{x} \sum_{z=y+1}^{D-x} \frac{D!}{y!(x-y)!z!(D-x-z)!} \phi_i^y \right.$$
$$\left. \cdot [\lambda(x)]_i^{x-y} \cdot \phi_{\bar{i}}^z \cdot [\lambda(D-x)]_{\bar{i}}^{D-x-z} \right].$$

Note that $\Pr\{N_i|N_i\}$ can be obtained from either equation by setting $d = 0$.

The effect of having more than one death per disturbance cycle is to reduce persistence time and hasten community turnover. This is because it takes fewer time steps to eliminate a species or make it a monodominant if larger changes per step are possible. Clearly, any species for which $N_i \leq D$ has a finite chance of going extinct in one time step. In the ergodic community, equilibrial relative abundances remain the same as in the case of $D = 1$, but they are approached faster, and the equilibrium variance about the mean is greater.

In the absorbing case ($m = 0$) with $D > 1$, the time to fixation (local extinction or complete dominance) is inversely proportional to the death rate. Figure 4.12 shows the linearity of the relationship for the case in which the initial abundance of the ith species is half the community size, $N_i(0) = J/2$. This is the case in which the mean time

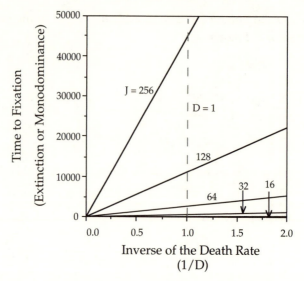

FIG. 4.12. Time to fixation (local extinction or monodominance) in the absorbing case of zero-sum ecological drift for death rates $D > 1$, for a species with an initial abundance of $J/2$. This is the abundance with the longest time to fixation. Time to fixation is inversely proportional to the death rate, D. A doubling of community size results in approximately a fourfold increase in time to fixation. Fixation times for the $D = 1$ case are indicated by the vertical dotted line.

to fixation is maximal (because the starting abundance is equidistant from the two absorbing states). A doubling of community size leads to an approximately fourfold increase in the time to fixation. Because of the rapid increase in time to fixation with increasing community size, it is useful to replot figure 4.12 on a semilogarithmic scale (fig. 4.13). This reveals the shorter times to fixation that occur as the death rate is elevated above unity. The ratio of fixation or absorption times asymptotically approaches the ratio of the death rates for large community size, J:

$$\frac{T_{D=1}}{T_{D=D'>1}} \to \frac{D'}{1} = D' \quad \text{as} \quad J \to \infty.$$

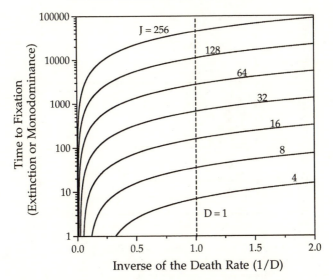

FIG. 4.13. Time to fixation (local extinction or monodominance) in the absorbing case of zero-sum ecological drift, as a function of the death rate and the size of the community. The relationship between fixation time and the inverse of death rate is linear, but the fixation time axis has been log-transformed to reveal the family of curves for different community sizes. Fixation times for the $D = 1$ case discussed earlier are indicated by the vertical cross-sectional line.

Small communities deviate by going to fixation faster than predicted from the asymptotic expectation (i.e., $T_{D=1}/T_{D=D'>1} > D'$), as shown in fig. 4.14. The asymptotic behavior for large J arises because the dynamics of the random walk of the ith species are approximately binomial in large communities. Conversely, when community size is small, the dynamics are more noticeably affected by the hypergeometric sampling without replacement in the death process, which hastens fixation. The shortening of the expected time to fixation is accompanied by a large reduction in the frequency of very long persistence times. Recall that times to fixation are approximately gamma distributed. The gamma distribution is a density function with most of

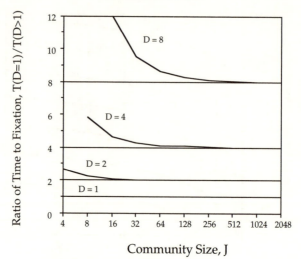

FIG. 4.14. Shortening of the time to fixation (extinction or monodominance) in the absorbing case of a community undergoing zero-sum ecological drift under a disturbance regime of $D > 1$. The y-axis is the ratio of the time to fixation for $D = 1$, to the time to fixation, for $D = D'$, where $D' = 2, 4$, and 8 are illustrated. The ratio of fixation times asymptotically approaches D' for large community size, J. Ratios deviate from D' at small community sizes due to the influence of hypergeometric death process. Curves are the ratios of fixation times for an initial abundance of $J/2$.

its probability density concentrated at low fixation times, relative to the range of possible fixation times. Therefore, for example, a twofold reduction in mean fixation time implies a major change in the shape of the distribution. Figure 4.15 illustrates the qualitatively large change in the distribution even for a very small community ($J = 8$) when the mortality is increased from $D = 1$ to $D = 4$ individuals per disturbance. The change in the distribution of extinction times is important and is often overlooked when the focus is strictly on the mean time to fixation. One of the important conclusions is that increasing the magnitude of the disturbance not only shortens the mean time to extinction, but also greatly

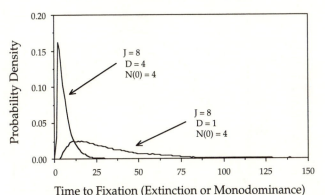

FIG. 4.15. Gamma-like distributions of the time to fixation (local extinction or monodominance) in the absorbing case of a small community ($J = 8$) undergoing zero-sum ecological drift, showing that the dramatic contraction of the tail of long fixation times with an increase in the mortality rate is increased from $D = 1$ to $D = 4$ per disturbance cycle.

curtails the chances for long-term survival through drifting relative abundance.

In discussing the effects of disturbance rates greater than $D = 1$, I have so far concentrated on the time to fixation in the absorbing case. This is because the main effect of $D > 1$ is to hasten fixation in the absorbing case, and to hasten attainment of equilibrium in the ergodic case. The expected relative abundance of the ith species in the ergodic community is unaffected by $D > 1$: the expected abundance remains JP_i. However, the variance in relative abundance increases modestly for larger D. Figure 4.16 gives a numerical example for a community of size $J = 64$, and for an immigration rate half the mortality per disturbance cycle. Even given a 32-fold difference in death rates ($D = 1$ to $D = 32$), there is a maximal increase of only about 20% in the standard deviation of relative abundance, which occurs at $P_i = 0.5$.

109

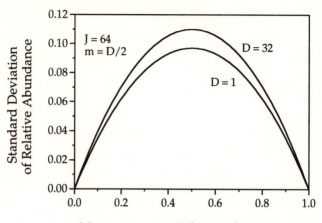

FIG. 4.16. Effect of increasing the death rate on the standard deviation of the relative species abundance of the ith species in an ergodic local community undergoing zero-sum ecological drift. The case of $J = 64$ is illustrated. The two curves compare $D = 1$ (lower curve) with $D = 32$ (upper curve), representing a wide range of disturbance regimes.

The deterministic analog to the stochastic dynamical equations of ecological drift can also be written down. The deterministic rate of change of the abundance of the ith species is given by

$$\frac{dN_i}{dt} = \frac{(J - N_i)}{J}\left[mP_i + (1 - m)\left(\frac{N_i}{J - 1}\right)\right]$$

$$- \frac{N_i}{J}\left[m(1 - P_i) + (1 - m)\left(\frac{J - N_i}{J - 1}\right)\right].$$

Setting the derivative equal to zero, we find that the equilibrium abundance N^* of the ith species is: $N_i^* = JP_i$. This is identical to the expectation under ecological drift, where N_i^* is the equilibrium numerical abundance of the ith species in a community of J total individuals. Note, however, that there

110

is no variance about this abundance in the deterministic case. Evaluating the stability of the equilibrium, we find that the eigenvalue of the linearized equation about N_i^* is

$$\left(\frac{\partial(dN_i/dt)}{\partial N_i}\right)_{N^*} = m(1 - J),$$

which is always negative for community size $J > 1$; hence, the equilibrium is always stable.

I now turn to consideration of the second half of the theory, namely, the theory for metacommunity dynamics. When the two halves of the theory are put together, we will unify the theories of island biogeography and relative species abundance into a single quantitative theory, and in so doing, obtain a complete analytical solution to the classical island-mainland problem posed by MacArthur and Wilson more than 30 years ago.

SUMMARY

1. This chapter develops the first half of the unified neutral theory and describes the dynamics of an arbitrary species undergoing zero-sum ecological drift in a local community as a function of local community size, the immigration rate, and the abundance of the species in the metacommunity.

2. The time to extinction or monodominance of the species in the local community can be very long if the community is large and the focal species is reasonably abundant initially. However, these times are also a function of the disturbance rate, and they are generally much shorter if the disturbance rate of the community is high.

3. The expected abundance of an arbitrary species in the local community is independent of the immigration rate and depends only on local community size and the

111

abundance of the species in the metacommunity. However, the variance in abundance is also a function of the immigration rate.

4. When local communities are very isolated from the metacommunity (low rate of immigration), the probability density functions of abundance are U-shaped, and the species spends most of its time either absent from the local community (the usual case) or occasionally monodominant.

5. When local communities are strongly coupled dynamically to the metacommunity by high immigration, relative abundances in the local community are more similar to those in the metacommunity.

6. Variances in local abundance are predicted to be a parabolic function of metacommunity relative abundance. However, detecting this curvilinearity may be difficult in samples when the commonest species is still a small fraction of the metacommunity, in which case the theory predicts a log-log linear relationship between variance and mean abundance.

Metacommunity Dynamics and the Unified Theory

According to MacArthur and Wilson's theory of island biogeography, a given local community or island achieves a steady-state species richness under a persistent rain of immigrants of already extant species inhabiting the much larger metacommunity source area. In the previous chapter I derived the probability density function for the abundance of the ith species in an ergodic local community as a function of P_i, the metacommunity relative abundance of the ith species. The ergodic community and MacArthur and Wilson's theory both implicitly assume the permanence of the ith species in the metacommunity. In reality, of course, communities are only ergodic on local spatial and temporal scales. Sooner or later, every species in the metacommunity suffers a final global extinction. Thus, the dynamics of particular species in metacommunities are governed by absorbing, not ergodic, processes, albeit with very slow dynamics due to the stabilizing effect of the law of large numbers.

Because all species ultimately go extinct, diversity is maintained, in the last analysis, solely by the origination of new species in the metacommunity. This is true whether or not niche assembly tends to stabilize communities on small spatio-temporal scales. MacArthur and Wilson's theory is conceptually incomplete in this regard because it lacks a speciation mechanism. Although species can go extinct in their model, no brand-new species are allowed to arise in their source area or on their islands. Speciation in the metacommunity is the analog of immigration in the theory of island biogeography. As I will endeavor to prove in this chapter, incorporating speciation into the theory of island biogeography has some surprising and potentially far-reaching implications. One of the most important is that it unexpectedly results in unification of the theories of island biogeography and relative species abundance. The unified theory reveals the existence of a fundamental biodiversity number, θ, that controls not only species richness but also relative species abundance in the source-area metacommunity.

In the absence of a generally accepted, quantitative, genetical, or ecological theory of speciation (Stebbins 1950, Mayr 1963, Rosenzweig 1978, White 1978, Templeton 1981, Barigozzi 1982, Singh 1989), I have chosen to model speciation by the simplest possible mechanism. I leave the details of species origination vague and simply erect a parameter for the probability of speciation—however one may choose to define species and speciation. New species arise in the theory like rare point mutations, and they may spread and become more abundant, or more often, die out quickly. I now introduce a parameter, ν (appropriately pronounced "nu"), for the speciation rate, defined as the probability of a speciation event per birth in the metacommunity. I make no assumptions about the tempo of speciation other than to surmise that it is likely to occur at an extremely slow rate.

This mode may reasonably characterize many speciation events, such as the origin of new plant species by abrupt changes in ploidy number (Stebbins 1950, Arnold 1997). To give it a name for present purposes, I will dub it the *point mutation mode* of speciation. However, many species arise through the vicariant allopatric subdivision of ancestral species (Mayr 1963) and never pass though a period of absolute rarity at origination. In the present chapter, I restrict myself to the analytically more tractable case of speciation as if by point mutation. Allopatric speciation is examined in chapter 8, where I also study a mode of speciation in which species arise by the random partition of a preexisting species into two daughter species. I will call this mode the *random fission mode* of speciation. Random fission captures the essence of allopatric speciation for purposes of the present theory, in which the actual physical nature of the barrier causing the allopatry is assumed to be unimportant. As I will show in chapter 8, the point mutation and random fission modes of speciation have different consequences for the distribution of metacommunity species richness and relative species abundance. Also one more parameter is needed to uniquely determine the metacommunity under random fission than under point mutation. Random fission speciation has a strong effect on mean times to extinction as well as steady-state metacommunity species richness and relative species abundance.

For the moment, however, let us restrict the discussion to the point mutation mode of speciation. Including speciation in the theory creates special theoretical difficulties to overcome. Our most familiar theories in ecology all concern the population dynamics and community ecology of specific, named, or labeled species, each of which has an assigned dynamical equation or set of equations. However, because the dynamics of any given set of species in the metacommunity obey an absorbing process (all species eventually go extinct), no fixed, nontrivial equilibrium dominance-diversity distribution can exist for any set of named species in the metacommunity. Thus, the analysis of metacommunity dynamics is qualitatively different from most classical ecological theory. Note that it is also different from the theory employed to describe the dynamics of the ith species in the ergodic local community in the last chapter. Nevertheless, a metacommunity equilibrium between speciation and extinction under zero-sum ecological drift does exist. A steady-state species richness and dominance-diversity distribution will arise in the pool of unnamed species slowly turning over in the metacommunity. This steady-state abundance distribution of unnamed transient species in the metacommunity is what we must now derive.

Before delving into this theory and its predictions, it is useful to illustrate briefly the empirical patterns of metacommunity relative species abundance for which we seek a theory. In the first figure of this book (fig. 1.1), I illustrated the dominance-diversity curves for a rather heterogeneous array of communities. However, here I show a more homogeneous set of communities—the relative abundance of tree species in closed-canopy forests—for historical and peda-

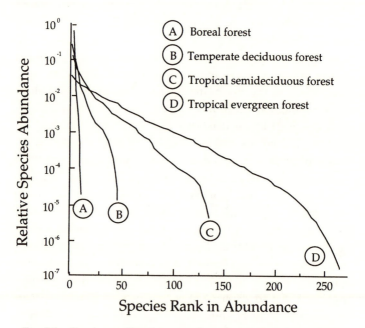

FIG. 5.1. Dominance-diversity curves for tree species in four closed-canopy forests, spanning a large latitudinal gradient. The four curves seem to represent a single family of mathematical functions, suggesting that a simple theory with few parameters might capture the essential metacommunity patterns of relative species abundance in closed-canopy forests. Redrawn from Hubbell (1979).

gogical reasons. I choose them because they are the communities with which I am most familiar and because their dominance-diversity curves are the very ones that stimulated the central theoretical idea in this book. Four dominance-diversity curves for tree communities spanning a latitudinal gradient are shown in figure 5.1. They range from a boreal coniferous forest to a lowland evergreen forest in the equatorial Amazon region.

The boreal forest is a 0.2 ha sample of the red spruce–Fraser fir community located on the summit of Clingman's Dome in Great Smoky Mountains National Park (sampled before the advent of acid rain) (Hubbell, unpublished data). This forest displays the geometric progression of relative species abundance characteristic of simple tree communities having high dominance and few (<10) species. This forest is similar to communities that Motomura (1932) described and modeled, and whose relative abundance patterns were later attributed to niche-preemption by Whittaker (1965) (chapter 2). Next there is a distribution for a species-rich, low-elevation temperate forest, a 1 ha sample of cove forest, also in Great Smoky Mountains National Park (Hubbell, unpublished data). This sample has about forty species and exhibits a typical S-shaped lognormal-like distribution of the sort found by Preston (1948). Finally, there are two tropical forests, one relatively species-poor (about 120 species in 13 ha of semideciduous forest in Guanacaste, Costa Rica) (Hubbell 1979), and the other a 4 ha sample of a species-rich (>200 species) evergreen forest near Belém, Brazil) (Hubbell 1979). Both of these tropical tree communities, although represented by different sample sizes, exhibit lognormal-like distributions of relative species abundance. The differences in these dominance-diversity curves are real; they do not result from the different plot sizes.

What is especially intriguing about these dominance-diversity curves is not their differences, but their similarities. The four curves appear to form a single family of closely related functions. This strongly hints that a simple dynamical theory with a parsimonious number of parameters might exist that predicts them all and shows how they are mathematically related. Three of the curves are S-shaped and appear lognormal-like but with different means, variances, and numbers of species. As species richness decreases, the distribution of relative species abundance becomes steeper, and the common species become even more dominant. Thus, there is an increase in the variance of relative species abundance as communities become poorer in species.

The only dominance-diversity curve that would appear not to be lognormal-like is the distribution for the boreal forest. But perhaps this interpretation is incorrect. The smooth progression of the dominance-diversity curves suggests that the boreal forest distribution is just an extreme case of a single process that produces lognormal-like distributions when there are many species, but geometric-like distributions when there are few species. With few species and an especially high variance in relative abundance, the distribution will appear relatively straightened and compressed on a semilog dominance-diversity plot compared to curves for species-rich communities.

Although each of the forests in figure 5.1 is a local community, the patterns in dominance and diversity are qualitatively representative of diversity patterns in boreal, temperate, and tropical forests, and of broad latitudinal changes in metacommunity relative abundance patterns. In island biogeography theory, local community dominance-diversity patterns are expected to be derivative of the metacommunity patterns. Therefore, I first consider the question of the origin and maintenance of species richness, dominance, and diversity in metacommunities. As I will now show, a pattern of dominance-diversity like that in figure 5.1 arises as a consequence of metacommunity dynamics under zero-sum ecological drift. This is the second of the theorems that follow from the biotic saturation of ecological communities (chapter 3).

Let us now turn to the problem of the simultaneous zero-sum ecological drift of all the species in the metacommunity. This problem is harder than the one we considered in chapter 4, where we only had to keep track of the dynamics of a single species at a time. An analogous multinomial problem exists in the neutral theory of population genetics, enabling us to take advantage of an analytical strategy developed by Ewens (1972) and Karlin and McGregor (1972) to solve for the metacommunity steady-state dominance-diversity distribution. Let J_M be the size of the metacommunity and let ν be the speciation rate. Now consider a discrete-time, nonoverlapping-generations model of metacommunity dynamics. We will assume that the metacommunity is so large that sampling with replacement is a good approximation (cf. chapter 4) so that multinomial probabilities accurately describe processes of ecological drift in the metacommunity undergoing zero-sum dynamics. For simplicity, let a single individual be capable of reproduction. Let n_i^t be the abundance of species i in generation t. Then, under ecological drift, the multinomial probability that the abundance of the ith previously extant species will be n_i^{t+1} in generation $t+1$, $i = 1, 2, 3, \ldots$, plus abundance x_k of the kth newly arisen species, $k = 1, 2, 3, \ldots$, is

$$\Pr\{n_1^{t+1}, n_2^{t+1}, \ldots, x_1, x_2, \ldots, n_1^t, n_2^t, \ldots\}$$
$$= \frac{J_M!}{\prod_i n_i^{t+1}! \prod_k x_k!} \prod_i \left[\frac{n_i^t(1 - \nu)}{J_M} \right]^{n_i^{t+1}} \cdot \prod_k \nu^{x_k}.$$

This multinomial probability implies that, in principle, a specifiable Markovian process exists for the multinomial zero-sum random walk. However, the process is obviously extremely complex, and there are a very large number of states to consider for reasonably large J_M. Therefore, an

119

alternative analytical strategy is needed to find the stationary dominance-diversity distribution in the metacommunity.

The strategy consists of calculating the unconditional probability as $t \to \infty$ of every possible configuration of relative species abundance in a sample of J individuals drawn at random from the metacommunity. These configurations range from the low-diversity extreme of a single species having J individuals, to the high-diversity extreme of J separate species, each represented by a single individual—and all possible configurations in between. At equilibrium, by definition there is no change in the expected abundances of unlabeled species represented by $1, 2, 3, \ldots$ individuals from one generation to the next. We can make use of this fact to solve the problem. Although the number of probabilities to calculate is still large to very large, it turns out to be possible to write down a fast sampling algorithm for the computation of the equilibrium dominance-diversity curve, the recipe for which is given in chapter 9.

Let us first write down a function for the probability that two individuals drawn at random from the metacommunity in generation $t + 1$ are of the same species, as a function of the same probability in generation t. If the two individuals are to be of the same species, neither individual can have just speciated, the probability of which is $(1 - \nu)^2$. Both individuals can either have had the same parent (with probability $1/J_M$) in the preceding generation, or they could have been offspring of different parents of the same species, which in turn shared a common ancestor in some earlier generation. Let F_2^{t+1} be the probability of drawing two individuals of the same species in generation $t+1$. Since all individuals of a given species can trace their ancestry back to a common ancestor (the original speciation event), we can write the following recursive function:

$$F_2^{t+1} = (1 - \nu)^2 \left[\frac{1}{J_M} + \left\{ 1 - \frac{1}{J_M} \right\} F_2^t \right].$$

When the metacommunity dominance-diversity equilibrium is reached, then there will be no change in this probability from one generation to the next, i.e., $F_2^{t+1} = F_2^t = F_2$. Solving for F_2, and ignoring small and higher-order terms in ν, which we can safely do since the speciation rate ν is a very small number relative to unity (e.g., $\nu \ll 10^{-10}$) we have

$$F_2 = \frac{(1-\nu)^2}{J_M - (1-\nu)^2(J_M - 1)} \cong \frac{1}{1 + 2J_M\nu}.$$

It turns out that $2J_M\nu$ is a composite parameter that appears throughout the subsequent theory. Because of its fundamental importance, we will give this parameter a special symbol, θ. Thus, $\theta = 2J_M\nu$, and $F_2 \cong (1 + \theta)^{-1}$. Although J_M is a very large number, and ν is a very small number, the product θ is finite and of intermediate value.

Now consider the probability of drawing three individuals of the same species at random from the metacommunity. None of the three can be a new species [probability $(1-\nu)^3$]. Furthermore, there are now more combinatorial ways of drawing the three individuals of the same species than in the case of two individuals. All three could be offspring of the same parent in generation t, an event with probability $(1/J_M)^2$. Or they can have descended from two parents, with probability $3(J_M - 1)(1/J_M)^2$. Finally, they could have descended from three different parents of the same species, with probability $(J_M - 1)(J_M - 2)(1/J_M)^2$. This gives the recursive equation

$$F_3^{t+1} = (1 - \nu)^3(1/J_M)^2\big[1 + 3(J_M - 1)F_2^t$$
$$+ (J_M - 1)(J_M - 2)F_3^t\big].$$

Again, at equilibrium there is no change in this probability between generations t and $t + 1$, so that $F_3^{t+1} = F_3^t = F_3$. Substituting in the expression for F_2 and solving for F_3, and again ignoring small and higher-order terms in ν, we obtain

$$F_3 \cong 2!(2 + 2J_M\nu)^{-1}F_2 = \frac{2!}{(1 + \theta)(2 + \theta)}.$$

again where $\theta = 2J_M\nu$. By induction, one finds that the probability of drawing J individuals all of the same species is very close to

$$F_J \cong \frac{(J - 1)!}{(1 + \theta)(2 + \theta)\ldots(J - 1 + \theta)}.$$

But this is also the probability that a random sample of J individuals contains just one species. We now need to compute the probability of each multispecies configuration among the J individuals in the sample. So, for example, we can now use F_J to find the probability that a random sample of J individuals contains $J - 1$ individuals of one species, and one individual of a second species. For any single ordering of individuals, the probability is $F_{J-1} - F_J$. But in this instance there are J possible orderings of $\{J - 1, 1\}$ individuals in the sample, so the probability is therefore

$$\Pr\{J - 1, 1\} = J\{F_{J-1} - F_J\} \cong \frac{J(J - 2)!\theta}{(1 + \theta)(2 + \theta)\ldots(J - 1 + \theta)}$$

for $J \geq 3$; and for $J = 2$, the probability is $\Pr\{1, 1\} = \theta/(1 + \theta)$.

Continuing in this manner for other multispecies configurations, we can obtain by induction (Karlin and McGregor 1972) the desired unconditional probability for an arbitrary dominance-diversity configuration. For a sample of size J individuals, the probability of obtaining S species with $n_1, n_2, \ldots, n_S$ individuals, respectively, where $J = \sum n_i$, is

$$\Pr\{S, n_1, n_2, \ldots, n_S\} =$$

$$\frac{J!\theta^S}{1^{\phi_1}2^{\phi_2}\ldots J^{\phi_J}\phi_1!\phi_2!\ldots\phi_J!\prod_{k=1}^{J}(\theta + k - 1)}$$

where $\theta = 2J_M\nu$ and ϕ_i is the number of species that have i individuals in the sample of size J. This proves that under ecological drift there is a nontrivial dominance-diversity equilibrium between speciation and extinction among the unlabeled species in the metacommunity.

However, we are still one step away from having the expected equilibrium dominance-diversity distribution. With no loss in generality, rank order the species in each dominance-diversity configuration, whose probability, $\Pr\{S, n_1, n_2, \ldots, n_S\}$, we have just computed, from the commonest (first position) to the rarest (last position). Let a given ranked dominance-diversity configuration be denoted by $\{r_1, r_2, \ldots, r_S, 0, 0, \ldots, 0\}$, where sufficient zeros are added to fill out the set $\{r_i\}$ to a total of J elements. (The set is filled out to J, because this is the maximum number of species that can occur in a sample of size J.) Adding zeros to the set does not change the corresponding probability, $\Pr\{S, r_1, r_2, \ldots, r_S, 0, 0, \ldots, 0\}$.

It is now possible to write down the expected abundance r_i of the ith ranked species in the equilibrium metacommunity dominance-diversity distribution for a sample of size J, as follows:

$$E\{r_i | J\} = \sum_{k=1}^{C} r_i(k) \cdot \Pr\{S, r_1, r_2, \ldots, r_S, 0, 0, \ldots, 0\}_k.$$

where C is the total number of configurations, $r_i(k)$ is the abundance of the ith ranked species in the kth configuration, and $\Pr\{S, r_1, r_2, \ldots, r_S, 0, 0, \ldots, 0\}_k$ is the probability of the kth configuration. The rank order distribution of species abundances, i.e., the metacommunity dominance-diversity curve, is now simply the ordered expectations, $E\{r_i\}, i = 1, 2, \ldots$, ordered such that species of the lowest rank are the commonest.

From these results we can compute the expected species richness and relative species abundance in a metacommunity obeying zero-sum ecological drift at equilibrium between speciation and extinction. The preceding two equations are functions of a single parameter θ (apart from sample size J). Since these equations completely determine the metacommunity diversity equilibrium, we have a truly remarkable result: *θ controls not only the equilibrium species richness but also the equilibrium relative species abundance in the metacommunity.* Parameter θ is a dimensionless, fundamental quantity that appears pervasively throughout the remainder of the theory at all spatio-temporal scales. For this reason, I believe that θ is justifiably named the *fundamental biodiversity number.*

How θ controls the shape of the dominance-diversity curve in the metacommunity is shown in figure 5.2. When θ is small (e.g., 0.1), the expected dominance-diversity curve is steep and geometric-like, with high dominance. However, as θ becomes larger, the expected dominance-diversity distribution becomes more lognormal-like, exhibiting the typical S-shaped curve observed in many species-rich communities (e.g., figure 5.1). At infinite diversity, in the limit as $\theta \to \infty$, every individual sampled represents a new and different species, regardless of how large a sample is taken. In this limiting case, the dominance-diversity curve becomes a perfectly horizontal line. At the other extreme, when $\theta = 0$, the distribution collapses to a single monodominant species throughout the metacommunity.

As the sample size increases toward infinity, the expected dominance-diversity curve rapidly converges on a stable relative species abundance distribution for a given θ. The distributions for a sample size of 100,000 are close to the limiting distributions, and are shown in figure 5.3, in which the equilibrium relative abundances are plotted as percentages.

Anticipating a result of chapter 6, it turns out that *the fundamental biodiversity number θ is asymptotically identical to Fisher's α*—the measure of diversity in the logseries distribution of Fisher et al. (1943) originally proposed more than 50 years ago. Moreover, *the zero-sum multinomial distribution of relative*

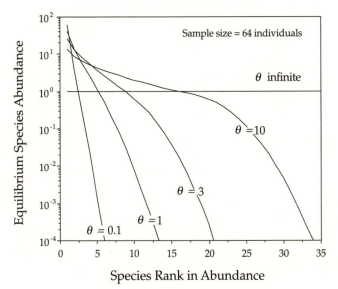

FIG. 5.2. Expected metacommunity dominance-diversity distributions for a sample of 64 individuals, for various values of the parameter, θ. When θ is small, the expected dominance-diversity curve is geometric-like. As θ becomes larger, the expected dominance-diversity curve becomes lognormal-like. As $\theta \to \infty$, the distribution approaches a horizontal line. In the limit, when θ is infinite, every individual in the sample is a new and different species, however large a sample is taken.

species abundance in the metacommunity is asymptotically identical (in the infinite size limit) to the logseries! Watterson (1976) obtained a similar result for the equivalent genetical theory of Ewens (1972) and Karlin and McGregor (1972). Thus, we finally have a dynamical theory for diversity based on fundamental birth and death processes that justifies the empirically fit logseries distribution of Fisher et al. (1943).

Recall, however, that Preston (1948) rejected the logseries because it always predicted that there would be more rare species than he observed in his samples of relative species abundance (see chapter 2). When Preston plotted the frequency of species in doubling classes of abundance, in

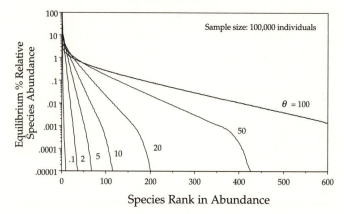

FIG. 5.3. As sample size $J \to \infty$, the distribution of relative species abundance approaches a stable limiting distribution of relative abundances for a given value of θ, which is estimated here from a sample of 100,000 individuals.

large samples there was almost always an interior mode in one of the abundance octaves representing species having more than a single individual. It was this interior mode which generated the characteristic lognormal-like appearance of the distribution. In contrast, when Fisher's logseries was similarly plotted, the category of singleton species always had the most species. Preston had no theoretical explanation for his empirical generalization.

It turns out that Fisher and Preston were both correct but on very different spatial and temporal scales. According to the unified theory, *an interior mode and a lognormal-like distribution (i.e., a zero-sum multinomial) of relative species abundance is expected in a local community or island, whereas a logseries-like distribution is expected in the metacommunity.* This is the theoretical expectation only under the point mutation model of speciation, however. Under random fission speciation, the metacommunity distribution is *also* a zero-sum multinomial, as we shall discover in chapter 8. But for the moment, let us explore why the theory predicts zero-sum multinomial distributions for local communities but logseries-like distributions for the metacommunity under point-mutation speciation. This requires that we now formally unify the theory of biogeography and relative species abundance over the two spatio-temporal scales.

To unify the theory, let us derive the dominance-diversity distribution that arises on an island experiencing immigration from the metacommunity—the classical island-mainland problem posed by MacArthur and Wilson (1967), extended to relative species abundance. Since we have already derived the rank-order abundance distribution for species in the metacommunity, then for the ith ranked species in the metacommunity, we have

$$P_i = E\{r_i\}/J_M = \sum_{k=1}^{C} r_i(k) \cdot \Pr\{S, r_1, r_2, \ldots, r_S, 0, 0, \ldots, 0\}_k/J_M.$$

From this we see that it is no longer necessary to specify P_i because $E\{r_i\}$ is completely determined by the fundamental biodiversity number, θ. Hence, we can dispense entirely with any and all species-specific parameters in the theory. *The unified theory then simplifies to just three parameters: the fundamental biodiversity number θ, the probability of immigration m, and the local community size J.* The parameter θ is of course a compound parameter of the metacommunity size J_M and the speciation rate ν, but these two parameters appear combined in θ in the theory for point mutation speciation. However, under random fission speciation, it will be necessary to keep these two parameters separate, and then four parameters need to be specified to uniquely determine the meta-community (chapter 8). To calculate the probability density function for the ith species in the local community, we simply substitute $E\{r_i\}/J_M$ for P_i in the eigenvector of equilibrium abundances derived in chapter 4.

By itself, the eigenvector for the ith species is not sufficient to obtain the multispecies dominance-diversity curve

127

for the local community or island. This requires calculating the equilibrium of the joint random walk, under zero-sum ecological drift, of all species in the local community, *and* coupled by immigration to the metacommunity. Consider a local community of size $J < J_M$, which is semi-isolated by dispersal limitation from the metacommunity ($m < 1$). Suppose that there are S species as determined by θ in the metacommunity of size J_M. Create a transition probability matrix whose states are all possible combinations of integer abundances of the S species that sum to the local community size J. Consider the analytically more tractable case in which only a single death and replacement occurs ($D = 1$). In this case, transition probabilities are nonzero only between abundance states that differ by one substitution, or states that remain unchanged. Thus, the probability that species i increases by one individual and species j decreases by one individual is given simply by

$$\Pr\{N_i + 1, N_j - 1, N_k, \ldots, N_S | N_i, N_j, N_k, \ldots, N_S\}$$
$$= \frac{N_j}{J}\left[mP_i + (1 - m)\left(\frac{N_i}{J - 1}\right)\right].$$

and the probability of no change in relative species abundance is

$$\Pr\{N_i, N_j, N_k, \ldots, N_S | N_i, N_j, N_k, \ldots, N_S\}$$
$$= \sum_{i=1}^{S} \frac{N_i}{J}\left[mP_i + (1 - m)\left(\frac{N_i - 1}{J - 1}\right)\right].$$

The eigenvector of the matrix of these transition probabilities gives the equilibrium probabilities of each relative abundance combination for the local community distribution. Let this eigenvector be denoted by $\phi(\kappa)$, where κ is the index for each relative abundance combination. Then in strictly analogous fashion to the computation of the dominance-diversity distribution for the metacommunity, the expected abundance of the ranked species in the local community of size J is given by

$$E_{local}\{r_i|J\} = \sum_{\kappa=1}^{C} r_i(\kappa) \cdot \phi(\kappa),$$

where C is the total number of configurations, $r_i(\kappa)$ is the abundance of the ith ranked species in the κth configuration, and $\phi(\kappa)$ is the probability of the κth configuration. As in the metacommunity case, the abundances of ranked species, i.e., the local community dominance-diversity curve, is now simply the ordered expectations, $E\{r_i\}$, $i = 1, 2, \ldots$, ordered such that species of the lowest rank are the commonest. It is important to note in the local community case, unlike the metacommunity, that r_i in the above equation does not refer to the rank of the labeled ith *species* in the metacommunity equilibrium distribution. It refers instead to the ith rank *position* in the local community, which can be occupied at any given moment by *any* of the metacommunity species. Thus, the preceding equation computes the expected abundance of the rank 1, 2, 3, et seq. positions regardless of which species currently occupies that rank in the local community. This will be made clearer by following a worked numerical example that is given in chapter 9 for illustration.

The computation of $\phi(\kappa)$ and of the preceding equation is unfortunately only feasible analytically up to a J of about 10, after which the number of combinations of relative abundance makes the analytical approach impractical. Fortunately, the equilibrium distribution of relative species abundance in the local community can be found very simply by simulation even for J values many orders of magnitude larger. The local community distributions of relative species abundance in the remaining part of this chapter are simulation results, but they have all been checked against the analytical results for small J_M and J.

This completes the unification of the theory on the local and metacommunity scales and completely solves the classical island-mainland problem of MacArthur and Wilson. We explore the implications of this unification next.

The neutral theory combining the metacommunity and local community dynamics into a single unified theory predicts a new statistical distribution of relative species abundance, the *zero-sum multinomial*, already alluded to in chapter 3. This distribution describes the relative abundance of species in the local community at dispersal equilibrium with the metacommunity. This distribution is sufficiently close to the lognormal to be easily confused with it, at least in its shape for common species. However, it differs from the lognormal in usually having a long tail of very rare species. An example of a zero-sum multinomial, plotted as a Preston-type curve, is shown in figure 5.4. The zero-sum multino-

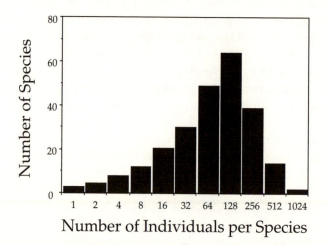

FIG. 5.4. An example of a zero-sum multinomial distribution, showing its typical skewing with a long tail of very rare species. This is the distribution predicted by the unified neutral theory for the distribution of relative species abundance in a local community at an immigration-extinction steady state with the metacommunity source area.

mial is also much more flexible in shape, and its shape is a function not only of the fundamental biodiversity number θ, but also of local community size J and the immigration rate m. Unfortunately, no analytical moment generating function for the zero-sum multinomial distribution is yet available, but the distribution is easily obtained by simulation recipes outlined in chapter 9.

We can also examine the shape of the zero-sum multinomial as a Whittaker-style dominance-diversity curve. When the immigration rate is greater than zero, a nontrivial equilibrium dominance-diversity distribution arises in the local community or island as a steady state between immigration and local extinction—just as MacArthur and Wilson's theory predicted a nontrivial island equilibrium in species richness. Consider a model island community of size $J = 1600$ experiencing different immigration rates from the metacommunity source area. We obtain a family of equilibrium island dominance-diversity distributions whose shapes depend upon the immigration rate (fig. 5.5).

When all of the deaths due to disturbance are replaced by immigrants, the dominance-diversity curve is the same as that of the metacommunity (i.e., logseries-like), with identical relative abundances as expected for a sample of size J individuals from the metacommunity. However, as fewer local deaths are replaced by immigrant individuals, the equilibrium dominance-diversity curve for the local community becomes steeper and more geometric-like. Note that it takes rather severe isolation to produce extremely steep dominance-diversity distributions—unless the metacommunity itself is species poor. For example, the steepest of the curves in figure 5.5 corresponds to just a single immigrant per 10,000 deaths. Conversely, when only 10% of the deaths are replaced by immigrants, the equilibrium dominance-diversity curve is still very similar to the source area distribution. All of these local distributions are examples of the zero-sum multinomial distribution. Thus, the uni-

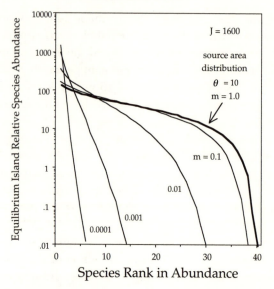

FIG. 5.5. Equilibrium island dominance-diversity distributions for an island under different rates of immigration from a mainland source area. Island community size has been set at 1600 individuals in this numerical example. The bold line is the expected dominance-diversity distribution if all deaths are replaced by immigrants from the source area. The equilibrium distribution becomes steeper and more geometric-like with increasing isolation from the metacommunity.

fied theory predicts that the local community distribution of relative species abundances will differentiate in both shape and in total species richness from the mainland metacommunity distribution. We can now state a general principle: *on islands, common species will be commoner, and rare species rarer, than in the metacommunity.* The degree of this differentiation is determined by the degree of isolation of the local community from the mainland, which in turn is controlled by the immigration rate m.

Herein lies the explanation for why Fisher and Preston were both correct. If we replot the relative species abundance distributions for the local community as Preston-type frequency plots, it is clear why (fig. 5.6). Consider once again a local community of size $J = 1600$. At infinite dispersal ($m = 1$), the local community is not isolated from the metacommunity. In this limiting case, the local relative abundance distribution will be a random sample of the metacommunity logseries-like distribution, and singleton species will be the most frequent abundance class in a Preston-type plot (panel A). However, as m becomes smaller, the island or local community becomes progressively more

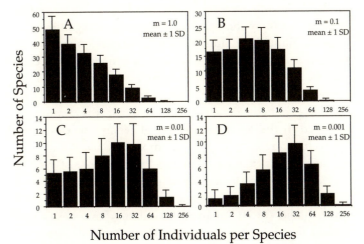

Number of Individuals per Species

FIG. 5.6. The effect of dispersal limitation (isolation) on the expected zero-sum multinomial distribution of relative species abundance in a model local community or island according to the unified neutral theory. Relative abundance distributions are plotted by doubling classes of abundance, following the method of Preston (1948). In all panels, the error bars represent ±1 standard deviation. (A)–(D): Model community of $J = 1600$ individuals and $\theta = 50$. (A) No dispersal limitation ($m = 1$). This is the distribution of relative species abundance expected in a random sample of 1600 individuals from the metacommunity logseries. (B) Relatively low local community isolation and dispersal limitation ($m = 0.1$). (C) Moderate isolation and dispersal limitation ($m = 0.01$). (D) Severe isolation and dispersal limitation ($m = 0.001$).

isolated, and the shape of the relative abundance distribution changes (panels B–D). As m decreases, rare species become rarer and less often present, and common species become commoner and more consistently present, in the local community. This results in a depression of the abundances of rare species and a rightward shift of the mode of the distribution to higher-abundance classes.

The unified theory thus predicts that *the shape of the local distribution of relative species abundance will be a function of the immigration rate*. This is the resolution of the conflict between Fisher and Preston. The unified theory explains the interior mode of the relative abundance distribution discovered by Preston as simply the result of restricted immigration, i.e., dispersal limitation. Thus, Preston's distribution was a sampling theory for relative species abundance in local communities, whereas Fisher's distribution, as we now realize in retrospect, was a sampling theory for the metacommunity.

These changes in shape also explain the asymmetry observed in the local distribution of relative species abundance in many Preston plots of empirical data. Recall from chapter 3, for example, that the Preston plots of relative species abundance in the 50 ha permanent plots in Pasoh Forest Reserve, Peninsular Malaysia, and on Barro Colorado Island (BCI), Panama, showed large excesses of rare species over what was predicted by the best-fit lognormal (see figs. 3.3, 3.4).

I have replotted the BCI and Pasoh distributions in figures 5.7 and 5.8, respectively, showing that the zero-sum multinomial distribution of the unified neutral theory fits the empirical distributions much better than the lognormal. The neutral theory achieves this improved fit with no more parameters than the lognormal (three). Probably more important than the number of parameters, however, is the fact that, for the first time, the parameters of the distribution are interpretable in terms of a birth-death-dispersal process. They aren't simply fitted means and variances of a

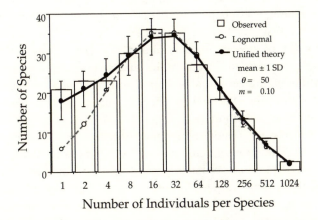

FIG. 5.7. Preston-type plot of relative species abundance for tree species >10 cm dbh in the 50 ha BCI plot, compared with expectations from the lognormal, and from the zero-sum multinomial of the unified neutral theory, for $\theta = 50$ and $m = 0.10$. The error bars are ±1 standard deviation.

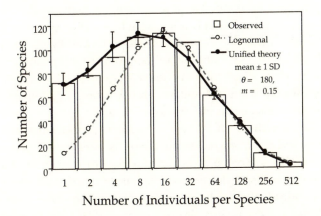

FIG. 5.8. Preston-type plot of relative species abundance for tree species >10 cm dbh in the 50 ha Pasoh plot, compared with expectations from the lognormal, and from the zero-sum multinomial of the unified neutral theory, for $\theta = 180$ and $m = 0.15$. The error bars are ±1 standard deviation.

135

generic statistical distribution. Moreover, because the theory generating the relative abundance distribution is dynamical, for the first time we can also put error bars on the expected number of species in each abundance class. These error bars arise from the demographic and community stochasticity underlying zero-sum ecological drift, and they are calculated from the expected variances in the steady-state relative abundance distribution.

Because the relative abundance distribution is a function of the immigration rate, we can use changes in shape of the distribution to estimate parameter m and thereby quantify the average dispersal limitation and degree of isolation affecting a given local community or island. For example, the estimated value of m for the BCI plot is 0.10. This is equivalent to the statement that 90% of the trees in the plot have been locally produced ("births"), and 10% were immigrants. This is a reasonable value for m given that 10% of the area of the plot lies within 17 m of the outer perimeter. The estimate of m for the Pasoh plot is slightly higher: 0.15. The main canopy at Pasoh is twice as high as at BCI (50 m vs. 25 m), which may mean that seeds disperse slightly farther in absolute distance from the average tree crown in the Pasoh forest. Once dispersal limitation is factored in, the expected equilibrium relative abundance distributions fit the observed distributions almost exactly (figs. 5.7, 5.8).

The goodness of fit of the theoretical distributions to observed dominance-diversity curves is often quite remarkable. This is a much more demanding fitting exercise than the Preston-type plot because the abundances of all species are individually displayed and ranked in the curve. I have plotted the dominance-diversity curves for BCI and Pasoh in figures 5.9 and 5.10, respectively.

The expected metacommunity logseries relative abundance distribution was calculated from a sample size of 1,000,000 (see fig. 5.3). It was then scaled to local community size for comparison to the local distribution in

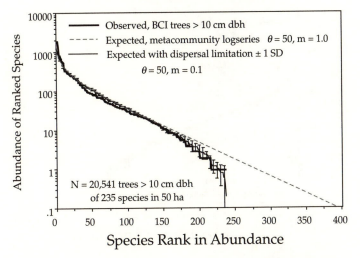

FIG. 5.9. Fitted and observed dominance-diversity distributions for trees >10 cm dbh in the 50 ha plot on Barro Colorado Island, Panama. The best fit θ had a value of 50. Note the departure of the metacommunity distribution for very rare species, but that the observed distribution is fit well once dispersal limitation ($m = 0.10$) is taken into account. The error bars are ±1 standard deviation.

each figure. The metacommunity logseries distribution is the diagonal line extending downward beyond the empirical curves to the lower right. The metacommunity distribution was calculated for a fitted θ value of 50 in the case of the BCI forest, and for a fitted θ value of 180 in the Pasoh forest. Then the parameters for dispersal limitation and local community size were included to predict the local community dominance-diversity curves in each forest plot. Local community size was 20,541 trees > 10 cm dbh in the BCI plot, but it was 28% higher (26,331) in the Pasoh plot. The previously estimated values of m of 0.10 and 0.15 for BCI and Pasoh, respectively, were used. The precision of the predicted local dominance-diversity curves in each plot is readily apparent from figures 5.9 and 5.10. The expected distributions fit even the abundances of the rarest species

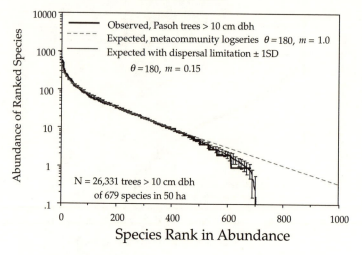

FIG. 5.10. Fitted and observed dominance-diversity distributions for trees > 10 cm dbh in the 50 ha plot in Pasoh Forest Reserve, Malaysia. The best fit θ had a value of 180. Note the departure of the metacommunity distribution for very rare species, but that the observed distribution is fit well once dispersal limitation ($m = 0.15$) is taken into account. The error bars are ± 1 standard deviation.

extremely well. The process of parameter estimation under the unified theory is discussed in detail in chapter 9.

The unified theory predicts that rare species will be even rarer and less frequent on islands or in local communities than one would expect from their metacommunity relative abundances. The theory explains this phenomenon as an interaction between dispersal limitation and extinction-proneness of rare species on islands. Rare species are more likely to go extinct per unit time, and are less likely to reimmigrate after local extinction, than common metacommunity species. The net result is that at equilibrium, *rare species will constitute a smaller fraction of the community on islands than in the mainland metacommunity.* This has important implications for conservation, because it means that rare species will be harder to conserve in a fragmented landscape.

Let us now consider a case that is more like the classical island-mainland case discussed by MacArthur and Wilson. Their theory hypothesized that the lower species richness on oceanic islands compared to same-sized sample areas on the mainland was due in part to lower rates of immigration than would occur among adjacent areas of a continuous mainland (chapter 1). The best test of the theory's predictions would be a case in which islands were once connected to a mainland but subsequently became isolated from it. If the immigration to an island is suddenly greatly reduced, the island will experience a steady loss in species richness, accompanied by a shift to more geometric-like, dominance-diversity distributions. Figure 5.11 illustrates the transient

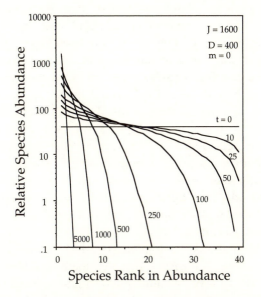

FIG. 5.11. Transient dynamics of the dominance-diversity curve in an island community of size $J = 1600$ undergoing zero-sum ecological drift under complete isolation from the metacommunity. Numbers indicate how many disturbance cycles since the island community was initiated with initiated with forty equally abundant species. The absolute death rate per disturbance is 400 in this example.

behavior expected during diversity loss in a model island community of size $J = 1600$ individuals that is completely isolated from the mainland metacommunity. For illustrative purposes, the initial community on the island was started with forty equally abundant species. In this case of complete isolation, the equilibrium community is a single monodominant species.

Such a test was possible in the Pearl Archipelago in the Bay of Panama (fig. 5.12). Several years ago, S. J. Wright enumerated tree species >10 cm dbh on three small islands (Chapera, Cocos, and Platanal) with similar topography, soils, and climate in the Pearl Archipelago in the Bay of Panama. During the Wisconsin glacial maximum, these islands were part of a broad, continuous coastal plain attached to the mainland. At that time they presumably had the full complement of mainland tree species. However, since isolation, they would have been expected to lose species through local extinction, and this appears to have been the case (fig. 5.13), as predicted by the theory of island biogeography. If the unified neutral theory applies as well, then the dominance-diversity distributions for the tree communities on these islands should also have become steeper and more geometric-like. This too is observed. The most abundant species reached levels of dominance more typical of a temperate deciduous forest than of a tropical forest. Each island ended up with a different dominant species in spite of similar topography, soils, and climate, also consistent with ecological drift.

The steepness of the dominance-diversity curves, and their nearly geometric form, suggests that these islands now experience low to very low immigration rates of trees. Suppose we can assume that these islands have reached their equilibrium dominance-diversity distributions. Then we can estimate the immigration probabilities for each island, assuming equilibrium, which are on the order of a low of six in 10,000 births for Cocos Island to a high of seven in 1000 births

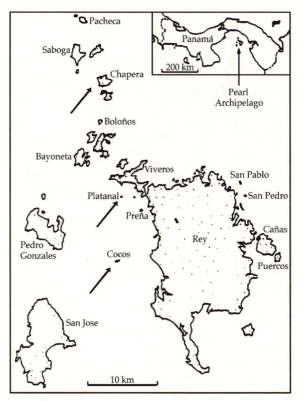

FIG. 5.12. Map of the Pearl Archipelago of islands off the southern coast of Panama. These islands were connected to the mainland during the Wisconsin glacial maximum, but became isolated about 10,000 years ago, when rising sea levels drowned the Pacific coastal plain. Tree communities were enumerated (trees > 10 cm dbh) on three islands—Chapera, Cocos, and Platanal—by S. J. Wright of the Smithsonian Tropical Research Institute.

for Platanal Island (Hubbell 1997). Cocos, the smallest and most remote island, had the steepest dominance-diversity curve and the fewest species, as expected. Platanal, the next-smallest island, nevertheless had the greatest species richness and the shallowest dominance-diversity curve. The proximity of Platanal Island to Rey Island, the largest of the

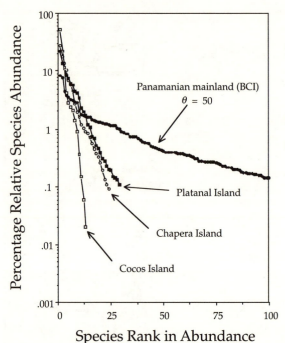

FIG. 5.13. The dominance-diversity distributions for the tree communities on Chapera, Cocos, and Platanal Islands of the Pearl Archipelago, compared to the dominance diversity for the forest on Barro Colorado Island (BCI), which serves as a representative Panamanian mainland site. The BCI data are truncated at one hundred species so that the dominance-diversity curves for the islands can be better displayed. Las Perlas data courtesy of S. J. Wright, Smithsonian Tropical Research Institute.

islands in the archipelago (fig. 5.12), may explain Platanal's higher diversity. Chapera, the largest of the three islands but also fairly remote from potential source areas, had intermediate tree species richness and dominance diversity. As predicted by theory, the variance in relative species abundance was greater on the islands than on the mainland. Dominance of the rank-1 species ranged from 29% on Platanal to 52% on Cocos Island.

If dispersal limitation in combination with local extinction is causing relative species abundances to differentiate on islands, a similar but weaker effect ought to be detectable in local stands of forest on continuous landscapes. In both the BCI and Pasoh plots we can, in fact, detect the predicted small shift in the shape of the dominance-diversity distributions in single hectares of forest relative to the 50 ha plot as a whole (fig. 5.14). These curves are compared on a percentage dominance basis to normalize the distributions for different total numbers of trees in 1 vs. 50 ha. As predicted, there is both a slight increase in dominance and a steepening of the dominance-diversity distribution in 1 ha plots vs. the entire 50 ha plot in both forests. In spite of major differences in species composition, stem density, tree species richness, and fitted θ values, the BCI and Pasoh forests are remarkably similar in the small-scale spatial differentiation

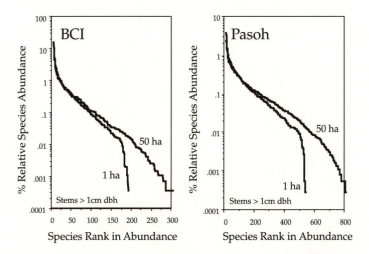

Fig. 5.14. Small-scale spatial differentiation of the dominance-diversity curve within the 50 ha plots on Barro Colorado Island, Panama (*left*), and in Pasoh Forest Reserve, Malaysia (*right*). The mean dominance-diversity curve for single hectares is shown along with the curve for the entire respective 50 ha plot.

of their dominance-diversity curves. The tailing-off of the rare species in the plots is as predicted from the greater effect of dispersal limitation on rare species.

We also find a similar differentiation of local and regional dominance-diversity curves on much larger spatial scales. For example, here are data from a study of the forests of Belize (Bird 1998). A series of 1 ha plots was established in closed-canopy forest in a 200 km north-south transect running most of the length of the country (fig. 5.15). We can use the total dataset as a measure of metacommunity biodiversity, and compare this with the biodiversity of the 1

FIG. 5.15. Map of Belize showing the locations of thirty-eight 1 ha plots of closed-canopy forest in a 200 km north-south transect. Data from Bird (1998).

ha plots. When we do this, we obtain a very good fit for the metacommunity with a θ of 66 (fig. 5.16). There is a slight tailing-off of rare species even in this dataset, indicating that there is dispersal limitation even at these very large geographical scales. However, there is much stronger evidence of dispersal limitation for the 1 ha plots, which show much greater dominance, fewer species, and steeper dominance-diversity curves, as predicted. The abundance of the rank-1 species in the 1 ha plots is about 15%, but it is only about 6% in the entire metacommunity.

Before concluding this chapter, it is worth taking a closer look at the fundamental biodiversity number θ. The

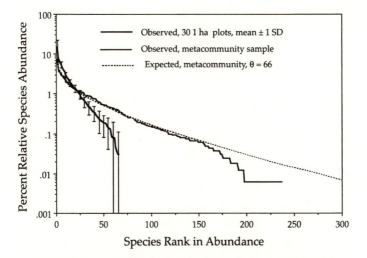

FIG. 5.16. The dominance-diversity curves for the forest of Belize. The pooled dataset for all thirty-eight 1 ha plots is quite well fit by a θ of 66. Nevertheless, even the metacommunity shows some evidence of dispersal limitation in the tailing off of the abundances of very rare species. The observed mean and standard deviations of the 1 ha plots show much greater dispersal limitation, and species richness is less than a third of the regional species richness. Data obtained from Bird (1998).

paramount importance of θ is that, for the first time, we have a formal theory of biogeography that ties speciation to patterns of diversity and relative species abundance on local to regional scales. Earlier I described latitudinal gradients in dominance-diversity patterns in closed-canopy forests (fig. 5.1). When θ is fitted to a geographical range of tree communities, values vary from a low of 0.15 for a sample of boreal spruce-fir forest in northern Newfoundland, to a high of 180 for Pasoh, a tropical lowland mixed-dipterocarp evergreen forest in Malaysia (table 5.1). The neutral theory explains these broad latitudinal patterns of dominance diversity in tree metacommunities as primarily the result of geographical variation in the fundamental biodiversity number θ.

Why is there geographic variation in θ? It could arise through systematic geographic variation in the size of metacommunities, in the speciation rate, or in some combination of the two. The metacommunity dominance-diversity equilibrium will also reflect variation in exogenous extinction rates that are additional to those expected under ecological drift. In high-latitude forests, speciation rates appear to be low relative to extinction rates, and geometric-like distributions obtain. Conversely, at equatorial latitudes, speciation rates are evidently high relative to extinction rates, and so species accumulate, resulting in very flat, lognormal-like steady-state distributions of relative species abundance. As I will discuss further in chapter 6, dispersal limitation also increases steady-state metacommunity diversity. The theory does not explain why extinction rates should be higher relative to speciation rates at high latitudes. One possibility is that disturbances are more frequent and/or more severe in high-latitude communities, elevating extinction rates relative to speciation rates. This was the thesis of the time-stability hypothesis put forward by Sanders (1968, 1969) and Slobodkin and Sanders (1969). They argued that the relatively greater age and stability of the tropics has allowed

TABLE 5.1. Estimated values of the fundamental biodiversity number θ for a broad range of closed-canopy tree communities, mainly New World

Forest Type	Forest Location	Biodiversity Number	Source of Data
Boreal forest	Flower's Cove, Newfoundland	0.15	Hubbell (unpubl.)
	Clingman's Dome, Great Smoky Mountains National Park, TN	0.22	Hubbell (1979)
	Mount Washington, NH (mid-elevation)	0.50	Braun (1950)
Northern hardwoods	Adirondacks, NY	2.0	Braun (1950)
Mixed temperate deciduous forest	Cumberland Plateau, KY	5.0	Braun (1950)
	Cover forest, Sugarlands, Great Smoky Mountains National Park, TN	7.1	Hubbell (1979)
Tropical semi-deciduous (dry) forest	Forest near Bagaces, Guanacaste, Costa Rica	24.0	Hubbell (1979)
Tropical semi-evergreen (moist) forest	Barro Colorado Island, Panama	50.0	Hubbell et al. (unpubl.)
Tropical evergreen forest	Pasoh Forest reserve, Negeri Sembilan, Peninsular Malaysia	180.0	Manokaran et al. (1992)

species to accumulate in the tropics and persist longer before extinction occurs. Another contributing factor is undoubtedly a simple climate-area effect. Terborgh (1973) has noted that there is far more land area per degree temperature change in the tropics than there is at higher latitudes.

I do not know how large realized θ values can become, but probably very large in microbial communities. Recent work inferring the diversity of soil microbial communities from surveys of DNA sequence variation for both culturable and nonculturable species suggests that, while diversity is not infinite, it is very large. John Tiedje (pers. comm.) found *no* repeats in more than four hundered strain isolates, i.e., a perfectly flat dominance diversity curve with every "species" represented by one and only one sample.

The fundamental diversity number θ is remarkable because it is finite and a moderately sized number. Its finiteness is remarkable because θ is the product of two numbers that differ enormously in size: the metacommunity size, J_M, and the speciation rate, ν (times two). We know very little about these numbers, but we can be certain that the speciation rate per birth is a very small number. Because the observed value of θ ranges roughly between 0.1 and 200, we must conclude that the metacommunity size J_M must be a very large number, i.e., on the order of 0.05 to 100 times the inverse of the speciation rate. For example, suppose the speciation rate is one in a trillion (10^{12}) births. Then the metacommunity size must be between roughly 50 billion ($5 \cdot 10^{10}$) and 10 trillion (10^{14}) individuals (births).

What is a reasonable size for the speciation rate or for the metacommunity? At this point it is difficult to say. The value of θ can be held constant if the size of the metacommunity is reduced but the speciation rate is increased to compensate. I will return to the question of estimating speciation rates in chapter 8 and discuss the implied magnitude of speciation rates in greater detail then.

Because of the difficulty of coping with such large numbers of individuals, it may be easier to use an area-based formula for θ rather than an individual-based formula. Note that θ can also be written as a simple linear function of the area A_M occupied by the metacommunity:

$$\theta = 2\rho A_M \nu,$$

where ρ is the mean density of individuals per unit area in the metacommunity. Note that this formula shows that the fundamental biodiversity number θ is directly tied to the biotic saturation of landscapes. Note also that ρ measures the productivity of a unit of area in landscapes to support organisms. Therefore, θ ties the biodiversity of the metacommunity directly to the productivity of the landscape.

The unified neutral theory thus reaches the simple but potentially profound conclusion that one can predict both the species richness and relative species abundance in a metacommunity undergoing zero-sum ecological drift by knowing just a few things. One needs to know the area and the average speciation rate in the biogeographic region, the density of organisms per unit area, and, finally (as I shall show in chapter 6), the mean dispersal rate of species over the landscape. With these few parameters, the neutral theory gives a complete description of the metacommunity under ecological drift, random dispersal, and random speciation. The unified theory appears capable of explaining many broad as well as specific patterns in relative species abundance in ecological communities from a few very simple neutral assumptions. The generality of the theory as applied to a wide variety of communities will be explored more fully in chapters 9 and 10.

One of the most important findings of the unified theory is that metacommunity dynamics are so slow under drift that ecological and evolutionary rates of change will become commensurate on large landscape scales. Indeed, this must be true because we can prove that a diversity equilibrium will inevitably arise between speciation and extinction in the metacommunity. It is largely this single fact that has made a unified theory of biogeography, speciation, and relative species abundance a reasonable goal to pursue.

SUMMARY

1. The unified neutral theory predicts the existence of a dimensionless biodiversity number θ, which is equal to twice the speciation rate times the metacommunity size. This number appears to be fundamental in the sense that it appears throughout the theory at all spatio-temporal scales.

2. Under the "point mutation" mode of speciation, a formula containing θ as the only parameter completely controls metacommunity species richness and relative species abundance at steady state between speciation and extinction.

3. With two additional parameters, immigration rate and local community (island) size, the neutral theory also predicts species richness and relative species abundance in local communities (islands).

4. This relative abundance distribution on islands (local communities) is a new statistical distribution called a zero-sum multinomial, whose shape depends on θ, the size of the local community (island), and the immigration rate. The distribution usually has a long tail of very rare species. It can be logseries-like, lognormal-like, or geometric-like, depending on the degree of isolation from the metacommunity source area.

5. On islands (local communities), common species are commoner and rare species are rarer and less numerous than expected from random samples of the metacommunity. This is because rare species are more

extinction prone and less likely to reimmigrate quickly after local extinction.

6. These results reconcile the classical dispute between Preston and Fisher over the shape of the relative abundance distribution. In retrospect it is clear that Fisher was describing the metacommunity sampling distribution, whereas Preston was describing the local community distribution.

CHAPTER SIX

The Unified Neutral Theory and Dynamical Species-Area Relationships

The rate at which species accumulate with increasing area surveyed—the species-area relationship—is perhaps the most basic and fundamental problem in biogeography. Yet on first consideration, a positive species-area relationship appears to be little more than a trivial corollary of the principle that Earth and its limiting resources are permanently and completely saturated with organisms. In an infinitely diverse world ($\theta = \infty$), the number of species would equal the number of individuals, and the number of species would therefore increase linearly with area. Short of infinite diversity, however, the biotic saturation of landscapes dictates that the number of species must accumulate more slowly than the number of individuals. The inevitability of a positive species-area relationship has led some ecologists to suggest that it is of little biological interest (e.g., Connor and McCoy 1979, Gilbert 1980, Connor et al. 1983). However, the species-area problem is far deeper than it might at first seem. One must explain why species-area relationships show strong and recurrent qualitative and quantitative patterns (Johnson and Raven 1973, Connor and Simberloff 1978, McGuinness 1984, Williamson 1988, Palmer and White 1994, Rosenzweig 1995). This chapter explores the unified theory's explanations of these recurrent patterns.

A debate over the relationship between species number and area has persisted for a long time. This controversy has been partly prolonged by failure to recognize that the species-area relationship obeys different scaling rules on different spatial scales (Palmer and White 1994, Rosenzweig 1995). But it has also been controversial because there is no theoretical foundation for species-area curves that derives from fundamental processes of population dynamics. Like most contemporary theories of relative species abundance, most current theories of species-area relationships are static and ad hoc sampling hypotheses. MacArthur and Wilson (1967) discussed the lognormal in relation to species-area curves in the first chapter of their monograph. However, they never actually formally connected their dynamical theory of island biogeography to species-area relationships (Williamson 1988). Later, May (1975) analyzed the expected accumulation of species with increasing area under a variety of statistical distributions, including the broken stick, logseries, and the lognormal. This approach presupposes that what matters most to the species-area relationship is the sampling of local species abundance. It ignores not only community dynamics but also dispersal limitation. If species are dispersal limited, then the expected local relative abundances of species will not be the same as random samples of the metacommunity (chapter 5). All species are dispersal limited on some spatial scale, and dispersal limitation becomes increasingly important on larger scales.

In hindsight, however, May (1975) made the almost prescient observation that "if relatively small samples are taken from some large and homogeneous area, the relation between sample area (or volume), A, and the number of species represented in the sample, $S(A)$, is likely to obey the logseries." The unified theory's explanation for this observation is that taking many scattered small samples will tend to overcome and compensate for dispersal limitation, and thus more closely approximate a random sample of the metacommunity with its nearly logseries equilibrium distribution of relative species abundance (chapter 5).

In previous chapters I have focused on the predictions of the neutral theory for patterns of species richness and relative abundance in the metacommunity and in local communities or on islands. However, the explicit linkage of the theory to biogeography on continuous landscapes has not yet been made because space has been treated implicitly. Indeed, the original theory of island biogeography did not treat space explicitly, either. Treating space implicitly is permissible when discussing migration from a mainland metacommunity to islands whose dynamics occur on very different spatial and temporal scales. But it is no longer acceptable in a general theory of biogeography on continuous landscapes—the theory needed to achieve deeper understanding of species-area relationships. Treating space explicitly also causes one very significant change in the neutral theory. Under explicit space, the metacommunity is no longer uniquely determined by the fundamental biodiversity number θ alone. It also becomes necessary to specify the mean rate of dispersal of individuals over the metacommunity landscape.

Some analytical progress on dynamical contact-process models of species-area relationships has recently been reported by Durrett and Levin (1996), and their results will be discussed shortly. But many questions about species-area relationships and the distribution of species diversity on continuous landscapes remain unsolved analytically. Fortunately, however, little is lost because the spatially explicit version of the unified theory is easily simulated in numerical experiments, and the analytical results of previous chapters provide a useful guide to interpreting the results of these simulations.

In this chapter I first discuss the theoretical species-individual sampling curve that arises in the metacommunity under the simplifying assumption of no dispersal limitation. This result leads to a neutral theoretical justification for Fisher's α. To examine species-area relationships and spatial patterns of species diversity under dispersal limitation, I study a spatially explicit version of ecological drift. I analyze the dynamical behavior of a metacommunity landscape "tiled" with identically sized local communities that exchange migrants with neighboring communities. I then discuss the steady-state species-area relationships that arise on this landscape as a function of θ, dispersal rate m, and local community of size J.

Before delving into the theory, it is useful to summarize briefly the major empirical patterns of species-area relationships and the current theories to explain them. For a thorough review of the subject, one should consult the book by Rosenzweig (1995). The empirical study of species-area relationships dates at least from H. C. Watson, who published the first known species-area curve for the vascular plants of Great Britain in 1859, the same year that *The Origin of Species* appeared (Williams 1964). Watson found a linear relationship between the logarithm of the number of species present and the logarithm of the area sampled, over areas ranging from a square mile to all of Great Britain (fig. 6.1). This is the most common pattern found on regional scales within relatively homogeneous landscapes (Gleason 1922, Williamson 1988, Rosenzweig 1995). The relationship is given by:

$$S = cA^z,$$

where S is the total number of species encountered in geographical area A, and c and z are fitted constants. This equation has come to be known as the Arrhenius species-area relationship after its most ardent champion (Arrhenius 1921), and is certainly now the most widely accepted relationship (Kilburn 1966, MacArthur and Wilson 1967, May 1975, Connor and McCoy 1979, Sugihara 1981). Note that if we let the constant $c = \rho^z$, where ρ is the mean density of individuals per unit area, then from our first principle we

155

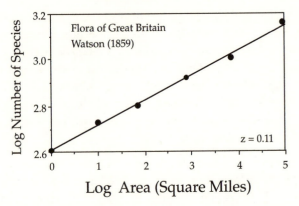

FIG. 6.1. Watson's species-area curve for vascular plants of Great Britain, accumulating species from a starting point in Surrey. After Williams (1964) and Rosenzweig (1995).

also obtain the number of species as a simple power function of the number of individuals J sampled $S = J^z$.

Recently, Harte et al. (1999) have noted that the Arrhenius relationship is a power law that implies self-similarity in the distribution and abundance of species. *Self-similarity* implies that if one takes the ratio of any two areas, as long as one maintains a constant area ratio, the ratio of the number of species will also be invariant. Harte et al. assumed that the Arrhenius power-law relationship holds at all spatial scales, from which they could deduce the distribution of relative species abundance that is implied by a particular slope or z value. It is very interesting that the shape of the relative species abundance distribution implied by this power law is not a lognormal, but a distribution with a long tail of rare species that is highly reminiscent of a zero-sum multinomial distribution (chapter 5). However, there is a major problem with this assumption. The problem is that self-similarity in the species-area relationship is not maintained at all spatial scales, which destroys any potential functional relationship between a particular slope z and the distribution of relative species abundance. This implies that there are different scaling rules underlying the species-area relationship on different spatial scales. Therefore, the conclusion of Harte et al. that a one-to-one mapping exists between the exponent of the Arrhenius species-area power law and the relative species abundance distribution is not correct (Borda de Agua et al., manuscript). We shall now explore the causes for why the Arrhenius relationship does not hold at all spatial scales.

In a review of the patterns and causes of species-area relationships, Williamson (1988) posed four questions about these patterns. First, why is the Arrhenius relationship so common? Second, why do some surveys clearly not fit the Arrhenius relationship? Third, why is z the parameter of the Arrhenius relationship generally between 0.15 and 0.40? Fourth, why is there so much variation in z among surveys? I will address each of these questions from the perspective of the neutral theory, but out of Williamson's original order.

Williamson's question number 2 has been empirically if not theoretically answered by Rosenzweig (1995). Rosenzweig found, in a review of a large number of species-area relationships, that the form of the species-area curve changes as a function of spatial scale. On local to global scales, species-area relationships are *triphasic*. The Arrhenius equation asserts that log-transforming both the number of species and area will linearize the species-area relationship, i.e., $\ln S = \ln c + z \cdot \ln A$. However, a linear log-log relationship is only typically obtained for intermediate spatial scales, but not on very local scales; and on large scales the slope increases. For example, Preston (1960) plotted the number of bird species over spatial scales ranging from an acre to the entire world, and obtained an S-shaped curve when log number of species was plotted against log area (fig. 6.2). In small contiguous areas, the curve was steep and curvilinear, then becoming shallower and log-log linear over intermediate to landscape and regional scales, and finally becoming

157

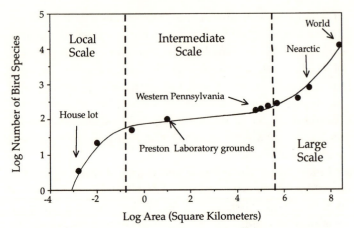

FIG. 6.2. Species-area curve for the world's avifauna, spanning spatial scales from less than one acre to the entire surface of the Earth. The S-shaped curve suggests that the sampling units change as area is increased, from individuals, to species ranges, and finally to different biogeographic realms at local, regional to subcontinental, and finally to intercontinental spatial scales. Data from Preston (1960).

steep once again over large intercontinental spatial scales, until the area of the entire world was included. The change in slope implies that scale-dependent changes in sampling units are occurring, a possibility of which Preston (1960) was clearly well aware. I will discuss this scale-dependent change in a more formal theoretical treatment shortly.

A similar S-shaped curve was obtained by Shmida and Wilson (1985), who plotted plant species-area relationships on local to global scales (fig. 6.3). They argued that the change in the form of the curve reflected changes in the biological determinants of plant species richness. On very local scales, niche-assembly rules would dominate. On somewhat larger spatial scales, mass effects and habitat diversity would become important. They define *mass effects* as an immigration subsidy from regional populations of a species that would go locally extinct without this immigration sub-

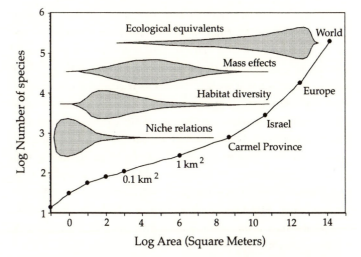

FIG. 6.3. Shmida and Wilson's (1985) hypothesis for the control of plant species diversity of different spatial scales, in relation to a species-area curve drawn from a local area in Israel (the mattoral region) to the entire world. Note the qualitatively similar shape to Preston's (1960) curve for the world's avifauna (fig. 6.2). Shmida and Wilson proposed that the biological determinants of species diversity changes from niche assembly on local scales, to habitat diversity and mass effects on intermediate scales, and to ecological equivalency on the largest spatial scales. The unified theory predicts these changes in the species-area relationship without positing any niche-assembly rules. Redrawn from Gaston (1994).

sidy. Finally, they propose that at very large scales there are many "ecologically substitutable" species. As I will demonstrate in this chapter, however, a completely neutral dynamical theory of species-area relationships is fully capable of explaining such triphasic species-area curves without invoking any niche-assembly rules whatsoever.

To classify these recurrent patterns, Rosenzweig (1995) offered a taxonomy of species-area curves in which he distinguished four types depending on the sampling scale and the evolutionary homogeneity or heterogeneity of the biota being sampled. Recasting his four types in the terminology of the unified theory, we have (1) species-area curves that arise in small areas within a single local community; (2) curves that arise in larger areas within a single metacommunity; (3) curves that arise among islands of an archipelago but still within the same regional metacommunity; and (4) curves that arise among different metacommunities with long-separate evolutionary histories. Rosenzweig's classification can be simplified under the unified neutral theory. All four species-area patterns can be explained as different manifestations of the same underlying dynamical process, namely, zero-sum ecological drift operating on different spatial scales. However, in this chapter I limit the discussion to species-area relationships on continuous landscapes and do not consider type 3 further.

Taking a dynamical perspective, one realizes that the species-area relationship must represent a steady-state phenomenon just like the steady-state species richness on islands in the theory of island biogeography. In chapter 5 we proved that, in the implicit spatial case for "point mutation" specation, there exists a fundamental number θ which determines a unique relative species abundance distribution in the metacommunity at equilibrium between speciation and extinction. Now, simply allow the identical process of zero-sum ecological drift with speciation to occur over a continuous regional landscape. Then a steady-state spatially distributed "standing wave" of species diversity must also exist over this landscape. Every species originates at some point or in some region on the landscape, disperses out from this point or region of origin, and ultimately goes extinct. Therefore, our task is to find the steady-state spatial distribution of species richness and relative abundance predicted by the neutral theory. We assume here that this equilibrium arises in a metacommunity undergoing zero-sum ecological drift, random dispersal, and random speciation on a large but finite, homogeneous landscape.

The theory's explanation for the triphasic nature of species-area curves can be explained qualitatively as follows (fig. 6.4). At very local spatial scales, the species-area curve is very sensitive to the local commonness and rarity of species, as individuals are collected one by one. However, on regional to subcontinental spatial scales, the rate of encounter of new species depends much less on relative species abundance, and more on rates of speciation, dispersal, and extinction, and the resulting steady-state geographic ranges of species. At very large, intercontinental scales, species accumulate faster again as major barriers to dispersal are crossed—barriers between different biogeographic realms with long separate evolutionary histo-

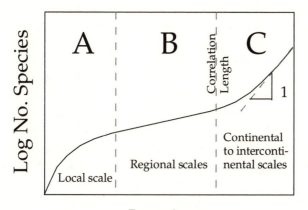

Fɪɢ. 6.4. The qualitative shape of the triphasic species-area curve, as predicted by the unified neutral theory. On local spatial scales (region A) the species accumulation curve is most sensitive to relative abundance of species in the local community. On regional spatial scales (region B), the species accumulation curve is less sensitive to relative species abundance and more to the encounter of the ranges of species at steady state between speciation, dispersal, and extinction. On very large spatial scales (region C), the correlation length of the biogeographic process has been exceeded, and sampling is of biogeographic realms with completely separate evolutionary histories. The correlation length defines the natural length scale of the biogeographic process.

161

ries (Rosenzweig 1995). There is an inflection in the species-area curve separating regional and very large scale sample areas, where the curve bends upward. This inflection point is the *correlation length* of the biogeographic process. The correlation length is very important because it specifies the *natural length scale* of the dynamical speciation-dispersal-extinction process, which in turn defines how big the natural biogeographic evolutionary units are on the metacommunity landscape. I will have more to say about biogeographic correlation lengths later.

According to the unified theory, the answer to Williamson's question number 4 is that slopes of species-area relationships vary so much because they depend dynamically on rates of speciation, dispersal, and extinction over regional to continental to intercontinental landscapes.

In general, sampling across dispersal barriers is always expected to steepen species-area curves because species that have not managed to cross the barriers will be relatively abruptly encountered for the first time. MacArthur and Wilson (1967) themselves invoked a verbal dynamical model to explain why species-area curves on the continuous mainland tend to exhibit shallower slopes than species-area curves for island archipelagos. This difference was attributed in part to the greater difficulty of dispersing to and among islands relative to dispersing among contiguous mainland communities, and in part to higher extinction rates on islands (MacArthur and Wilson 1967, Brown and Gibson 1983, Rosenzweig 1995). As we will show below, the unified theory predicts that the slope of the species-area curve will approach unity because at very large sampling scales, regional biogeographic processes become independent and uncorrelated.

I now turn to a more formal development of the neutral theory of species-area relationships. I first consider the theoretical species-area and species-individual curves for the metacommunity under the assumption of no dispersal limitation. Then I introduce dispersal limitation and determine how it affects the expected curves. Two of Williamson's questions still remain, namely: Why is the Arrhenius relationship so common (question 1). And, Why do the slopes (z values) of these relationships commonly lie between 0.15 and 0.4 (question 3)? I will offer the unified theory's explanations for these patterns in due course.

A dynamical theory of species-area relationships on all spatial scales is most easily approached through the theory for the metacommunity. We can relate the sample of size J to area by means of our first principle (chapter 3), which states that landscapes are essentially always biotically saturated with individuals of a specified community or taxon. Thus, $J = \rho A$, where A is the area occupied by J individuals and ρ is mean density. As noted in chapter 5, this relationship also holds for the full metacommunity, $J_M = \rho A_M$, so that we can express the fundamental biodiversity number θ as a simple linear function of area and density of organisms:

$$\theta = 2\rho A_M \nu.$$

This relationship implies that a function of θ exists which specifies both the species-individual and the species-area relationships. If dispersal is not limited, then the unified theory makes a simple prediction for the cumulative species-individual and species-area curves. We can easily derive the expected number of species S in a sample of J individuals chosen randomly from the metacommunity from the formula for the probability, $\Pr\{S, n_1, n_2, \ldots, n_S\}$ (see chapter 5). Thus, for $J = 1$, the expected number of species is clearly unity (from the formula, θ/θ). For $J = 2$, the probability of obtaining a single species with two individuals is $1/(\theta + 1)$, and the probability of two species each having a single individual is $\theta/(\theta + 1)$. Thus the expected number of species in a sample of two individuals is the sum of the two expectations: $E\{S|\theta, J = 2\} = E\{S = 1|\theta, J = 2\} + E\{S = 2|\theta, J = 2\}$, or

$$E\{S|\theta, J = 2\} = 1 \cdot \left(\frac{1}{\theta + 1}\right) + 2 \cdot \left(\frac{\theta}{\theta + 1}\right)$$

$$= \frac{\theta + 1}{\theta + 1} + \frac{\theta}{\theta + 1} = \frac{\theta}{\theta} + \frac{\theta}{\theta + 1}.$$

Continuing in this manner, one can show by induction that for arbitrary J,

$$E\{S|\theta, J\} = \frac{\theta}{\theta} + \frac{\theta}{\theta + 1} + \frac{\theta}{\theta + 2} + \cdots + \frac{\theta}{\theta + J - 1}.$$

This expectation is for the species-individual curve sampled at random from the equilibrium metacommunity obeying zero-sum ecological drift. It specifies the expected rate of addition of species as individuals are collected one by one, and in principle can be extended out to the metacommunity size, J_M. Note that we can convert this expectation into a species-area curve, $E\{S|\theta, \rho, A\} = \Sigma\theta/(\theta+\rho A-1)$ simply by making the variable substitution $J = \rho A$. This expectation was originally derived by Ewens (1972) for the sampling of neutral alleles in the "infinite allele" case, which is identical to the ecological drift case. If one applies this to sampling in a spatial context, however, it only applies under the limiting assumption of complete mixing of the species, i.e., no dispersal limitation, as we shall discuss in a moment.

We can now show, from the equation for the metacommunity species-individual curve, that the fundamental biodiversity number θ, which derives from the theory of zero-sum ecological drift, provides a justification for Fisher's α, the diversity parameter of the logseries distribution (Fisher et al. 1943). Recall from chapter 2 that Fisher's formula for species accumulation is

$$S(\alpha) = \alpha \cdot \ln\left(1 + \frac{J_M}{\alpha}\right).$$

As J_M increases, $S(\alpha)$ becomes asymptotic to $\alpha \cdot \ln(J_M/\alpha)$, a straight line on a graph of S vs. $\ln(J_M/\alpha)$ with slope α.

The theoretical species-individual curve predicted by θ and zero-sum ecological drift can be approximated closely by

$$S(\theta) \approx 1 + \theta \cdot \ln\left(1 + \frac{J_M - 1}{\theta}\right).$$

Note that as J_M increases, $S(\theta)$ also becomes asymptotically parallel (displaced by unity) to $\ln(J_M/\theta)$ with slope θ. Thus, we have a very important result, namely that, *as metacommunity size J_M, increases toward infinity. Fisher's α is asymptotically identical to the fundamental biodiversity number θ*:

$$\alpha \cong \theta = 2J_M\nu,$$

and the metacommunity distribution becomes asymptotically identical to Fisher's logseries. This result was also discovered for Ewens's formula in the infinite neutral allele case in population genetics (Watterson 1974). Because θ is not precisely equal to α, the distribution of relative species abundance in the metacommunity will be very similar to, but not precisely identical to, the logseries.

The implications of this simply stated result are quite profound. We have now connected Fisher's logseries distribution to the theory of metacommunities undergoing pure ecological drift. We therefore conclude that *if relative species abundances are well described by the logseries and Fisher's α, then these data are consistent with zero-sum ecological drift and with random sampling of the metacommunity.*

In the limiting case of no dispersal limitation, the metacommunity equation $E\{S|\theta, J\}$ gives the expected species-individual and species-area curves at all spatial scales, from local to global. It works on local scales when J is small, and on regional scales when J is large. Figure 6.5 illustrates

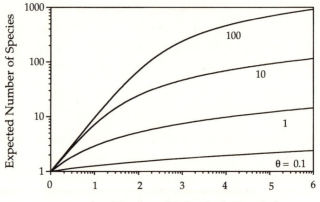

FIG. 6.5. Expected species-individual curves for values of the fundamental biodiversity number θ ranging over 3 orders of magnitude from 0.1 to 100. Note double log scales. These are expectations for a random sample of individuals from the metacommunity (no dispersal limitation). Individuals sampled can be converted to area by making the variable substitution, $J = \rho A$.

the expected metacommunity species-individual curves for three orders of magnitude of variation in θ on a double-log transformed plot. (Recall from above that we can convert cumulative individuals into cumulative area.) Note the curvilinearity of the lines on local scales, demonstrating the failure of the Arrhenius relationship, $S = cA^z$, on local scales (Williamson's question number 2). As mentioned above, this is because the species-area sampling process is especially sensitive to relative species abundance when J is small. At first, species are added fast with increasing sample size because the common species are collected quickly in the first samples, making the species accumulation curve rise steeply. Later, the curve rises more slowly as successively rarer and rarer species are collected. The theoretical curves never reach an asymptote, however. This is so because the infinite series $\sum_{J=1}^{\infty} \frac{\theta}{\theta+J-1}$ diverges as $J \to \infty$. Divergence is perfectly acceptable because an infinite number of species can be accommodated among an infinite number of individuals. Of course, real metacommunities are not infinitely large, although J_M is a very large number, and so, there can only be a finite number of possible species. Once a value of θ has been estimated, the expected number of species in a sample of any size can be determined, including, in principle, the entire metacommunity, assuming J_M can be reasonably estimated. Note also that because $\lim_{J \to \infty} \frac{\theta}{\theta+J-1} = 0$, successively rarer species are added at an ever decreasing rate as the number of individuals increases.

The curves in figure 6.5 also reveal the strong sensitivity of the species-individual and species-area curves to the fundamental biodiversity number θ. In a sample of a thousand individuals, for example, a community with a θ value of 100 will exhibit about twenty times as many species as a community with a θ value of unity, two orders of magnitude smaller. Because of the nonlinearity of the species-individual curves, however, two communities differing in their θ's do not maintain a constant ratio of species diversity as sample size varies. Note that as J becomes larger, the species-individual curves become more similar in slope and approach linearity on the double-log plot (fig. 6.5). This anticipates the log-log linear species-area relationships that the theory predicts on intermediate (regional) spatial scales.

The similarity of the theoretical curves in figure 6.5 to many empirically derived species-individual curves is readily apparent. Nearly all such curves are initially steep and then become shallower, whether the curves are plotted arithmetically or logarithmically. For example, figure 6.6 shows a sample of the species-accumulation curves reported by Sanders (1969) for benthic marine communities collected in transect trawls at four sites spanning a large latitudinal range. These curves were constructed from a series of small, scattered trawls and, therefore, as noted above, are expected to

167

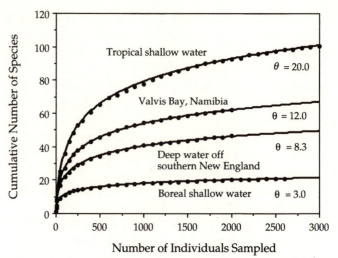

FIG. 6.6. Species-individual curves for benthic invertebrate conmmunities at four sites spanning a broad latitudinal range. Dots are Sanders's observations; solid lines are the expected curves. Redrawn from May (1975).

be closer to a random sample of the metacommunity. In fact, May (1975) found that these curves were well fit by Fisher's logseries, meaning that they are also expected to be fit by the unified theory's metacommunity distribution.

The theoretical metacommunity species-individual curve assumes random sampling, but it nevertheless also occasionally fits the data for local community species-individual curves quite precisely. A case in point is the excellent fit obtained for the species-rich genus of figs, *Ficus* (Moraceae), in the 50 ha plot on Barro Colorado Island, which yielded a fitted θ of 2.9 (fig. 6.7). The observed curve represents the mean of 10 randomized drawings of all individuals of the 13 fig species in the plot. The error bars are $\pm$ 1 standard deviation of species number for a given number of individuals drawn.

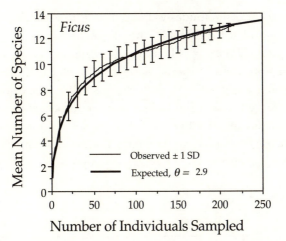

FIG. 6.7. Species-individual curve for 13 species of the genus *Ficus* in the 50 ha plot on Barro Colorado Island. The expected curve is a maximum likelihood estimated θ of 2.9. The error bars are ± 1 standard deviation of 10 random sampling of *Ficus* trees in the plot. The metacommunity distribution fits this genus well, probably because figs are very good dispersers.

When the metacommunity species-individual curve fits local data this well, it either means that the community or taxon is composed of species that are very good dispersers and are well mixed over fairly large landscape scales, or else that the sampling regime compensates to some degree for dispersal limitation. *Ficus* is a genus of stranglers or free-standing trees that are pollinated over relatively long distances by small windborne agionid wasps, and whose seeds are widely dispersed by fruit bats, frugivorous birds, and monkeys. This view of fig population structure is supported by studies of fig population genetics that indicate little genetic differentiation over large distances (Nason et al. 1996).

More generally, however, because of dispersal limitation, the metacommunity species-individual curve is not expected to fit local communities well, according to the unified theory. Consider, for example, the species-individual curve for the genus *Diospyros* (Ebenaceae), a species-rich genus of understory trees in the 50 ha plot at Pasoh Forest Reserve in Malaysia (fig. 6.8). These typically small-stature trees produce large multiseeded berries that are dispersed locally by ground mammals. After an initially rapid rise, the observed curve for *Diospyros* continues rising but more slowly than the theoretical curves, two of which are shown for comparison.

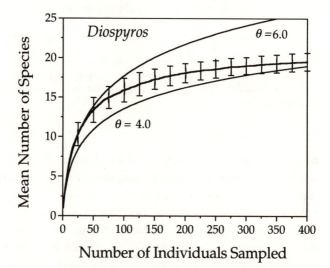

FIG. 6.8. Species-individual curve for 25 species of the genus *Diospyros* in the 50 ha plot in Pasoh Forest Reserve, Malaysia. Two theoretical curves for $\theta = 4$ and $\theta = 6$ are compared with the observed curve. The error bars are ± 1 standard deviation of 10 random samplings of *Diospyros* trees in the plot. The metacommunity distribution does not fit this genus because its species are poor dispersers. Increasing dispersal limitation on larger spatial scales causes the increasing lag in observed species accumulation as sample size increases, relative to the metacommunity curves that assume unlimited dispersal.

The local species-individual curve will rise more slowly under dispersal limitation than without it. Unfortunately, no analytical result of the effect of dispersal limitation on the species-individual curve is currently available. However, we can derive an approximation that appears to be quantitatively very close to the true function. In continuous space, note that the probability m of a local death being replaced by an immigrant must be a monotonically decreasing function of the number of individuals, J, occupying the given area, A. The probability of immigration is unity for $J = 1$ (chapter 4), and $m \rightarrow 0$ as $J \rightarrow \infty$. As noted previously, the species-individual curve approaches the power law $S = J^z$, which implies that the probability of immigration must decline due to dilution from increasing J in proportion to the inverse of J raised to some power, i.e.,

$$m(J) \approx J^{-\omega},$$

where the dispersal limitation parameter $\omega \geq 0$. Incorporating this function into the equation for the local species-individual curve, we obtain the following function for the rate of addition of species under dispersal limitation:

$$E[S|\theta, J] \approx \frac{\theta \cdot 1^{-\omega}}{\theta} + \frac{\theta \cdot 2^{-\omega}}{\theta + 1} + \cdots + \frac{\theta \cdot J^{-\omega}}{\theta + J - 1},$$

or that $E\{S|\theta, J\} \approx \sum_{i=1}^{J} \frac{\theta \cdot i^{-\omega}}{\theta + i - 1}$. Note that when $\omega = 0$, this equation reduces to the limiting metacommunity case of no dispersal limitation. It should be stressed that this equation is appropriate only for dispersal limitation on local spatial scales. On regional to continental spatial scales, the effect of dispersal limitation is actually to increase the slope of the species-area curve (see below).

I illustrate the goodness of fit of the species-individual function under dispersal limitation to the observed species-individual curve data for trees > 10 cm dbh in the 50 ha plot on Barro Colorado Island (fig. 6.9). In chapter 5 we

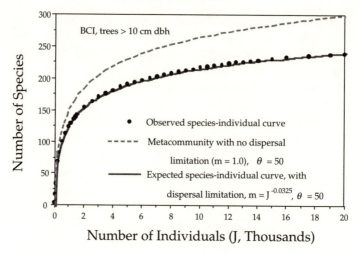

FIG. 6.9. Species-individual curve for trees > 10cm dbh in the 50 ha plot on Barro Colorado Island, Panama. The observed curve (dots) was obtained by averaging species richness in randomly chosen contiguous areas containing a given number of individual trees. The metacommunity species-individual curve overestimates the number of species in the plot by 65 species (28%). The parameters of the species-individual curve under dispersal limitation were estimated by maximum likilihood, and yielded $\theta = 50$ and $\omega = 0.0325$.

found that the dominance-diversity curve for BCI was well fit with a θ of 50 and a probability of immigration m of 0.10. The species-individual curve for BCI gives us a more precise method for measuring the probability of immigration, because it describes how species number increases at the level of the addition of individuals rather than of whole species. Moreover, we can now estimate the average community-level rate at which the probability of immigration declines as we increase the sample size, J, by estimating parameter ω. Figure 6.9 shows that the metacommunity species-individual curve considerably exceeds the observed number of species for a given number of individuals. It overestimates the true number of species in the 50 ha plot by more than a quarter (28% higher). However, if we take dispersal limitation into account ($m < 1$) and estimate the rate at which mean dispersal limitation increases with J, then we obtain a very good fit to the observed species-individual curve, with a maximum likelihood estimate for ω of 0.0325 for a θ once again of 50. Similar curves are obtained for the 50 ha plot in the Pasoh Forest Reserve. A more detailed discussion of species-area and species-individual curves in these plots can be found in Condit et al. (1996).

I now turn to regional and larger spatial scales, the domain in which Rosenzweig's (1995) species-area curve types (2) and (4) are found. As we move from local to regional to global spatial scales, dispersal limitation becomes far more important than relative abundance in controlling the species-area relationship. Thus the proportion of dispersal-limited species increases as area increases. Decreasing the average dispersal rate increases the proportion of species that will be dispersal limited for an area of a given size.

The neutral theory predicts opposite effects of dispersal limitation on the species-area curve on local versus regional scales. On local scales, dispersal limitation reduces the slope of the species-area curve because rare species are not encountered as fast as would be expected under random spatial mixing of species. On regional scales, dispersal limitation increases the slope of the species-area curve. This is not predicted by the theoretical metacommunity species-area curve, which exhibits an ever shallower slope with increasing area. However, this curve was derived assuming random sampling without dispersal limitation. The effect of dispersal limitation becomes increasingly pronounced as the sampling scale exceeds the steady-state range sizes of more and more species (i.e., exceeds the correlation length L), which in turn increases the encounter rate of previously untallied, dispersal-limited species. Therefore, *the unified theory's explanation for the change in the form of the species-area curve on different spatial scales is that the effect of dispersal limitation on the*

173

species accumulation rate changes from negative to positive with increasing spatial scale.

We can explore this change in the form of the species-area relationship more fully using two spatially explicit models of metacommunity drift on intermediate and large spatial scales. I first discuss the model of Durrett and Levin (1996), who studied a version of the "voter model" (Holley and Liggett 1975, Silvertown et al. 1992) applied to the species-area problem. The name "voter model" was originally inspired by theoretical studies of simple models of the spread of opinions. Imagine "voters" standing on an infinite gridded plain, with one person in each grid location. Each person keeps his or her opinion for an exponentially distributed time period, after which he or she adopts the opinion of a neighboring individual, such that the person located at (x) picks the opinion of the person at location (y) with probability $p(x, y)$. Durrett and Levin (1996) added mutation or "speciation" to this model, whose spatial dynamics evolves according to the following rules. Let the "species" occupying a given site be identified by some random number chosen on the interval $(0, 1)$. Rule 1 is that sites are always occupied by some species, but at rate δ, an individual dies and is replaced by an individual of a species chosen at random from the neighborhood, with probability $p(|y - x|)$. Rule 2 is that at rate ν, $\nu \ll \delta$, the occupant of a site changes or "speciates" into a new type $\in (0, 1)$ not seen before.

Durrett and Levin (1996) cite proofs in Bramson et al. (1996) of the following results for the special case of $p(|y - x|) = 1/4$ when $|y - x| = 1$, or that immigration is possible in one time step only from the 4 immediate neighbors (ignoring neighbors on the diagonals). They report approximations that become more exact as the speciation rate $\nu \to 0$. They relate their results to the correlation length L of the biogeographic process. In their model this distance is defined as the inverse of the square root of the speciation rate: $L = 1/\sqrt{\nu}$. This number is approximately equal to the expected distance under a two-dimensional random walk that an average species moves over its metacommunity lifetime from the site of its origination. They were able to prove several useful results, three of which will be mentioned here. First, they showed that the slope of the log-log species area curve approaches zero as the speciation rate $\nu \to 0$, and that when ν is small, this slope is approximately

$$\frac{\log S}{2 \log L} \approx \frac{2 \log \log L + \log(2/\pi)}{2 \log L} \to 0,$$

where S is the number of species. Note that $2 \log L$ is simply the logarithm of a square area whose side is equal to the correlation length L. Second, they showed that at length scales less than L, as $\nu \to 0$ the number of species at equilibrium approaches a constant. This is a weaker theorem than in the present theory of ecological drift, where one can prove that when $\nu = 0$, there can only be a single surviving species at equilibrium (chapter 4); but it agrees with the theory.

Durrett and Levin (1996) used their approximation for the slope of the log-log species area curve to provide a partial answer to Williamson's (1988) last question, which was: Why is there so much variation in the slope of species-area curves? If we insert various values of the speciation rate, substituting $1/\sqrt{\nu}$ for L, we obtain the relationship between speciation rate and z value shown in figure 6.10 for length scales less than L. The z value is low for slow rates of speciation and becomes progressively steeper for faster rates. Note also that the relationship is curvilinear: z values increase ever faster as the speciation rate increases.

Durrett and Levin (1996) showed that on length scales that are much larger than the correlation length L, the slope of the species-area curve becomes steeper and approaches unity. The slope increases to near unity because events become uncorrelated on length scales $\gg L$—spatial scales that are large relative to the equilibrium range sizes and

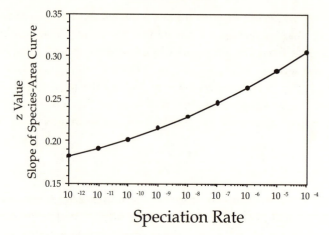

Speciation Rate

FIG. 6.10. Relationship between the z value (slope of the Arrhenius species-area curve) and the speciation rate, according to an approximation derived by Durrett and Levin (1996) for small v.

dispersal distances taken by species during their evolutionary lifetimes in the metacommunity. This explains the phenomenon of the steepening species-area curves discovered by Preston (1960) and Shmida and Wilson (1985) at large spatial scales (figs. 6.2 and 6.3). Rosenzweig's (1995) puzzlement over the fact that "the [species-area] curve among provinces has z values near unity" (p. 379) is also now explained. This effect is another reflection of the fact that there is extreme dispersal limitation between major biogeographic provinces.

Another useful result of Durrett and Levin's (1996) model is that it provides a theoretical justification for the prevalence of the log-log species-area relationship of Arrhenius (1921) on intermediate spatial scales—as does the present theory. This addresses Williamson's (1988) question number 1. Durrett and Levin's formula for the slope of the species-area curve is closely related to the unified theory's equation, $S = 1 + \theta \cdot \log[1 + (J-1)/\theta]$. To see this, observe that Durrett and Levin studied, in unified theory terms, a finite meta-community of size L^2 (since there was one individual per grid location), in which $J_M = \rho A = L^2$. For large metacommunities, $J_M >> \theta$, and, taking logs, $\log S \approx \log[\log(J_M) - \log \theta] + \log \theta \approx \log \log(J_M) + \log \theta = \log(2 \log L) + \log \theta = \log \log L + \log 2 + \log \theta$. Dividing both sides of the equation by $2 \log(L)$ to compute the slope of the species-area curve, we obtain

$$\frac{\log S}{\log(J_M)} \approx \frac{\log \log(J_M) + \log \theta}{\log(J_M)} = \frac{\log \log L + \log 2 + \log \theta}{2 \log L}$$

$$\approx \frac{\log S}{2 \log L}.$$

This formula clearly shows the influence of both J_M and θ on the slope of the approximately log-log species-area curve for large J_M or L. It differs somewhat from the formula of Durrett and Levin, which is expected since their formula was derived for a specific case of dispersal limitation (the voter model with $p(|y - x|) = 1/4$ when $|y - x| = 1$), whereas the unified neutral theory's formula is based on the metacommunity species-individual curve with no dispersal limitation. Using the above equation, we can also calculate how the slope of the metacommunity species-individual curve changes with the number of individuals sampled. The slope of the tangent at any J is approximately $[\log \log L + \log \theta] \log J$ for $J > 1$. From this expression, one could also construct the approximate species-individual curve for the metacommunity.

In contrast, the species-area slope calculation of Durrett and Levin cannot be used to compute the species-area curve for a given speciation rate. This is because in their formulation one cannot vary the area parameter—the square of the correlation length L—without simultaneously varying the speciation rate ν. A second "problem," more apparent than real, would seem to exist with the speciation rates that

Durrett and Levin (1996) needed to obtain z values similar to those of observed species-area curves (fig. 6.10). These speciation rates ranged from 10^{-4} to 10^{-12} per birth, which might seem to be unrealistically high.

These "problems" disappear entirely in the unified theory with the introduction of the fundamental biodiversity number θ, which applies equally well to the voter model with speciation studied by Durrett and Levin. Because θ is a product of J_M and ν, one can separate the effects of metacommunity size and speciation rate for the first time. A large metacommunity with a low speciation rate can have an identical species-area curve as a small metacommunity with a high speciation rate, so long as the fundamental biodiversity number θ remains the same (under point mutation speciation). Increasing the metacommunity size while holding the speciation rate constant will increase the slope of the equilibrium species-area curve, just as increasing the speciation rate will. Increasing J_M increases the number of speciation events per unit time because absolutely more births occur per unit time in a large metacommunity than in a small one.

I now study a somewhat more complicated model by simulation, one that is similar to the voter model, except that now we allow each grid location to be a local community of arbitrary size $J > 1$. Let each local community be coupled in a single time step through dispersal with probability m per birth with the local communities that comprise its immediate neighbors. The equations for the local dynamics of the ith species can be written down as follows. Let (x, y) be the coordinates and index name of the local community. Let deaths in local community (x, y) be replaced by local births or by immigrants with probability m in one time step from the surrounding frame of 8 neighboring communities, located at $(x - 1, y - 1)$ to $(x + 1, y + 1)$. This neighborhood is known as the "Moore neighborhood" in cellular automata theory. The metacommunity abundance parameter P_i is thereby replaced by the current abundance of species i summed over the surrounding frame of neighboring communities.

Including the probability of speciation, ν, the transition probabilities for the ith species having abundance $N_i(x, y)$ in the local community (x, y), with immigration from the immediate neighboring communities in one time step, then become

$$\Pr\{N_i(x,y)-1|N_i(x,y)\}=$$
$$\frac{N_i(x,y)}{J}\left[v+(1-v)\left\{m\left[1+\frac{N_i(x',y')}{J}-\sum_{x'=x-1}^{x+1}\sum_{y'=y-1}^{y+1}\frac{N_i(x',y')}{J}\right]\right.\right.$$
$$\left.\left.+(1-m)\left(\frac{J-N_i(x,y)}{J-1}\right)\right\}\right]$$

$$\Pr\{N_i(x,y)|N_i(x,y)\}=$$
$$\frac{N_i(x,y)}{J}(1-v)\left\{m\left[1-\frac{N_i(x,y)}{J}+\sum_{x'=x-1}^{x+1}\sum_{y'=y-1}^{y+1}\frac{N_i(x',y')}{J}\right]\right.$$
$$\left.+(1-m)\left[\frac{N_i(x,y)-1}{J-1}\right]\right\}+\frac{J-N_i(x,y)}{J}$$
$$\times\left[v+(1-v)\left\{m\left[1+\frac{N_i(x,y)}{J}-\sum_{x'=x-1}^{x+1}\sum_{y'=y-1}^{y+1}\frac{N_i(x',y')}{J}\right]\right.\right.$$
$$\left.\left.+(1-m)\left(\frac{J-N_i(x,y)-1}{J-1}\right)\right\}\right]$$

$$\Pr\{N_i(x,y)+1|N_i(x,y)\}=$$
$$\frac{J-N_i(x,y)}{J}(1-v)\left\{m\left[1-\frac{N_i(x,y)}{J}+\sum_{x'=x-1}^{x+1}\sum_{y'=y-1}^{y+1}\frac{N_i(x',y')}{J}\right]\right.$$
$$\left.+(1-m)\left[\frac{N_i(x,y)}{J-1}\right]\right\}.$$

For example, the first equation is the probability that species i in local community (x, y) will decrease by one individual in the next time step. For this to happen, an individual of the ith species in community (x, y) must die and not be replaced by another individual of species i. This replacement individual can be a new species (with probability ν) or be a preexisting species (with probability $1 - \nu$), in which case it can either be an immigrant or a local birth of some species other than i. Note that P_i is no longer a constant, as it was treated in chapter 4, but is replaced by the current relative abundance of the ith species in the frame of local communities surrounding the focal community (x, y). Also note that m is no longer the probability of immigration from the entire metacommunity, but only from the surrounding local communities.

Suppose the initial ancestral condition is a single, monodominant species whose individuals saturate the metacommunity plain and all local communities. Now start an engine of speciation, such that a new species arises somewhere on the plain with probability ν per birth. New species will arise at random points of origination on the finite but large landscape and, if they are lucky, will multiply and disperse before ultimately going extinct. Figure 6.11 is a cartoon illustrating the spatially explicit process of zero-sum ecological drift in the metacommunity. The figure shows four early stages in the evolution of a steady-state species diversity and its geographical dispersion over a small piece of the infinite plain. Initially, at $t = 0$ (panel A), the plain is occupied everywhere by a single, monodominant species (shading). At some later time, $t = t_1$ (panel B), a new species originates (white circle) from the original species. By time $t = t_2$ (panel C), the second species has increased in abundance and spread out from its point of origin. At any given moment each species will have a distribution that reflects its unique history of repeated colonizations and extinctions in local communities up to that point. Later, by time $t = t_3$ in this cartoon (panel D), a third species has arisen (black circle).

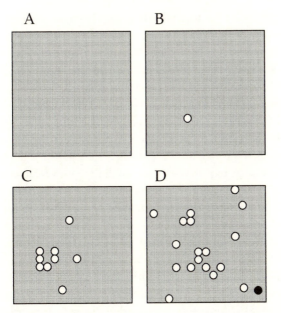

FIG. 6.11. Early stages in the evolution of species diversity on a portion of a homogeneous plain harboring a metacommunity. Initially (A), one species is monodominant in the metacommunity (shading). Later (B), a second species (white circles) originates, and subsequently (C) spreads out from its point of origination. Still later (D), a third species arises (black circle). The process continues indefinitely, and eventually a stochastic metacommunity steady-state species diversity between speciation and extinction is achieved.

Let us now study numerically the dynamical behavior of a metacommunity gridded into identical local communities, each of whose species obeys the above equations of zero-sum ecological drift. Conceptually, the metacommunity can be infinite in size, but in practice only finite representations are possible on a computer (and in the real world). One has basically two choices in modeling finite metacommunities: either with or without edges. There are advantages and disadvantages to both representations. To implement a metacommunity model without edges, one "wraps" opposite sides of the grid into a torus so that they become adjacent. The advantage of the torus model is that it completely eliminates edge communities and edge dynamics. However, it is unclear that species-area relationships will be the same on a torus and on a plain, particularly if the torus is relatively small. Moreover, to my knowledge there are no toroidal communities in nature, and their behavior is difficult to show in planar representations. One must disconnect the edges and "unwrap" the torus; but in so doing, the resulting maps make it appear that populations can suddenly jump from one side of the metacommunity to the other. This could introduce a significant artifact into the analysis of species-area relationships under dispersal limitation.

Of course, the disadvantage of planar models with edges is edge-effect dynamics. Local communities along edges behave differently because they have smaller source areas and more restricted immigration than local communities far from edges. The result is that edge communities are effectively more isolated and subject to ecological drift. They will generally exhibit steeper dominance-diversity curves than centrally located communities. Nevertheless, I prefer the planar approach, but one in which species-area relationships are studied far from edges to reduce edge-effect dynamics as much as possible. I simulated a relatively large metacommunity (101×101 local communities), but restricted my analysis of regional species-area relationships to a grid of 21×21 local communities at the center of the metacommunity, which comprises only 4.3% of the total metacommunity area. In the simulations that follow, I chose a local community size of $J = 16$ individuals unless otherwise indicated.

The first important result to establish is that a nontrivial metacommunity equilibrium species richness and relative species abundance distribution exists in the spatially explicit theory. We need to know that this equilibrium is unique and independent of initial conditions, and that it is similar to the equilibrium predicted by the metacommunity theory in chapter 5, when we treated space implicitly. These equilibria will not be identical, however, because dispersal limitation is present in the spatially explicit theory, whereas it is absent in the spatially implicit metacommunity theory. I am unable to prove uniqueness analytically in the spatially explicit model for local community size $J > 1$. However, simulation results leave no doubt that the equilibrium is nontrivial and unique, as is demanded by the spatially implicit theory given in chapter 5.

Figure 6.12 shows the convergence on the equilibrium metacommunity dominance-diversity curve starting from initial monodominance, for a case in which $D = 4(25\%$ mortality per disturbance cycle), $m = 0.5$, and $\theta = 10$. Figure 6.13 shows convergence to the same equilibrium from initial conditions of "infinite" diversity at the other

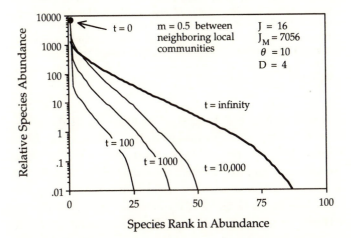

FIG. 6.12. Transient approach from initial conditions of monodominance to the equilibrium metacommunity dominance-diversity under dispersal limitation. Local community size $J = 16$. Metacommunity size $J_M = 7056$ individuals; $m = 0.5$; $\theta = 10$. The bold line is the equilibrium dominance-diversity curve ($t = 10{,}000$ birth-death cycles). The curves are shown only for the one hundred most common species. A stochastic equilibrium is achieved that is indistinguishable from that obtained under initial conditions of infinite diversity (cf. fig. 6.13).

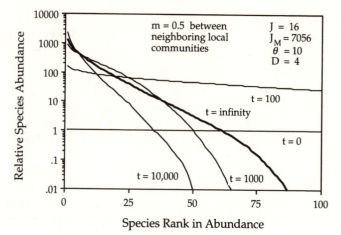

Fɪɢ. 6.13. Transient approach from initial conditions of "infinite" diversity to the equilibrium metacommunity dominance-diversity under dispersal limitation. Local community size $J = 16$. Metacommunity size $J_M = 7056$ individuals; $m = 0.5$; $\theta = 10$. The bold line is the equilibrium dominance-diversity curve ($t = 10,000$ birth-death cycles). The curves are shown only for the one hundred most common species. A stochastic equilibrium is eventually achieved that is indistinguishable from that obtained under initial conditions of monodominance (cf. fig. 6.12).

extreme, with one individual of each species at the outset. The curves represent means of one hundred simulations. In figure 6.13 the horizontal line at unity for $t = 0$ is the initial conditions of infinite diversity. The approach to equilibrium is more complex in this case than under monodominance (compare figs. 6.12 and 6.13). In the case of "infinite" initial diversity, it would appear from figure 6.13 that all species become more abundant early on, but in fact only some species do so. Figures 6.12 and 6.13 truncate the relative abundance distributions at the one hundred most common species.

Although I have illustrated only one numerical case in these two figures, the following conclusions are general and hold for any and all values of the theory's parameters in the spatially explicit unified theory. First, the equilibrium distribution of metacommunity diversity is independent of initial diversity conditions, as was the case in the spatially implicit theory. Second, the expected diversity distribution is uniquely determined by just three parameters: the fundamental biodiversity number θ, local community size J, and the probability of migration m. Finally, the approach to diversity equilibrium is much faster from initial conditions of infinite diversity (every individual is of a different species) than from initial conditions of monodominance.

This asymmetry stems from two factors. First, metacommunity species that are abundant and widespread are extremely resistant to displacement and extinction. Second, completely new metacommunity diversity must await speciation events, which are very infrequent, and even then the probability of successful establishment of new species is likely to be very low. This is particularly true under the "point mutation" model of speciation. However, if new species arise by "random fission," then successful establishment of new species is more likely. These results have important implications for conservation. When most species are rare, metacommunity diversity will be easy to lose, whereas once metacommunity diversity is lost, it will be much harder and slower to recover.

Having now established that a unique equilibrium dominance-diversity curve arises in the metacommunity in the spatially explicit unified theory, I return to the subject of species-area relationships. We are interested in the predictions of the spatially explicit model for species-area relationships on intermediate (regional) spatial scales—the scales on which the log-log linear Arrhenius (1921) relationship holds—and then on still larger spatial scales. Specifically, we wish to determine how variation in the fundamental biodiversity number θ and the dispersal rate m affect both the unit area intercepts (α diversity) and the slopes (z values, β diversity) of the equilibrium species-area curve at these spatial scales.

185

I examined steady-state species-area relationships on an intermediate spatial scale for a minimum area spanned by nine local communities to a maximum area representing 121 local communities each of size J = 4 sampled from the center of a metacommunity of 201 × 201 local communities, far from any edge. The initial conditions of the metacommunity were set to the dominance-diversity distribution dictated by a given θ for a metacommunity of size J_M = 201 × 201 × 4 individuals under no dispersal limitation. Species were assigned initial locations at random. I first established that 20,000 birth-death cycles with one death per cycle in each local community (5000 complete turnovers of the metacommunity) was more than adequate time for the species-area curves to equilibrate. Several typical sets of results are shown in figure 6.14. Each species-area curve in the figure represents a mean of one hundred simulations

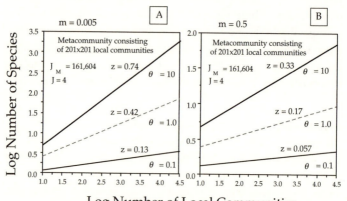

FIG. 6.14. Effect of varying the fundamental biodiversity number θ and the probability per birth of immigration m, into a local community from neighboring communities on the small-area intercept (α diversity) and the slope (z value, β diversity) of the species-area curve at regional spatial scales. Metacommunity consists of 201×201 local communities, each of size J = 4. (A) Immigration rate m = 0.005. (B) Immigration rate m = 0.5.

for a given θ and m. In the left panel, $m = 0.005$, a low rate of immigration, and in the right panel the probability of immigration is one hundred times greater ($m = 0.5$).

Several general conclusions are illustrated by these graphs that are consistent with the analytical results of chapter 4. First, the theory shows that log-log linear Arrhenius power law species-area curves are always obtained on intermediate spatial scales. Second, the intercept and the slope (z value) of the species-area curve are controlled by two parameters: the fundamental biodiversity number θ and the probability of immigration m. The effects of the two parameters differ. Increasing θ increases the z slope as well as the small-area intercept, raising both α and β diversity. Conversely, increasing the dispersal rate m *decreases* the slope of the species-area curve (lowering β diversity) but *increases* the small-area intercept (raising α diversity). The reasons for these contrasting effects are simple. When dispersal rates are rapid relative to speciation rates, z values are low and species-area curves are shallow. But when dispersal is very limited, then many new and local species will be encountered as sample area increases, and the species-area curve will have a steeper slope. Quantitatively, in this example (fig. 6.14) a 100-fold increase θ from 0.1 to 10 resulted in about a 5.7-fold increase in the z value, whereas a 100-fold increase in m from 0.005 to 0.5 resulted in a 55%–60% decrease in the z value. More generally, the ratio of increase or decrease will vary depending on the range of actual values of θ and m.

A large number of simulations show conclusively that the unified theory always generates log-log linear Arrhenius type species-area curves on intermediate spatial scales. Therefore, the more general question of how θ and m affect the z value of the species-area curve can now be answered. Figure 6.15 plots contours of equal slope z as a function of θ and m. In the figure, I have varied θ over slightly more than three orders of magnitude, from 0.1 to 200, spanning much of the range of θ's observed in nature (with the possible

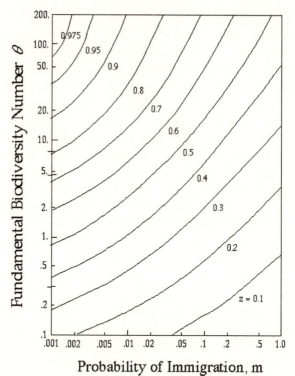

FIG. 6.15. The effect of varying the fundamental biodiversity number θ and the probability of immigration m on the slope (z)-value of the Arrhenius species area curve at regional spatial scales, according to the unified theory. Note that $z \rightarrow 1$ as $\theta \rightarrow \infty$ and that z becomes smaller as $m \rightarrow 1$. It should also be noted that $m = 1$ in the spatially explicit model does not correspond to infinite dispersal, as in the implicit-space theory, but rather to a case in which all local-community deaths are replaced by immigrants from the Moore neighborhood of immediately adjacent local communities. Results are from simulations (see text).

exception of some highly diverse microbial communities; see chapter 5). I have also varied immigration rate m over three orders of magnitude, from 0.001 to 1.0. The precise location of the contours of iso-z lines in $\theta - m$ space depends upon the exact nature of the dispersal function used in the model, but several qualitative conclusions are universally true.

First, a general result is that *predicted z values are steep for large θ and small m, and shallow for small θ and large m*. Second, the effect of an incremental change in the immigration or dispersal rate on the slope of the species-area curve is larger for large m than for small m. Note that as m decreases, the isoclines of equal z become shallower and more nearly parallel to the m-axis. Generally, for very low values of m (e.g., 10^{-5} or lower), changes in m have very little effect on the species-area slope and the dominant variable affecting z values is simply the fundamental biodiversity number $θ$. For a given $θ$, it requires a lower immigration rate to achieve a given z value when $θ$ is larger than when $θ$ is small. From this we conclude that *the slope of the species-area relationship becomes less sensitive to (more independent of) the dispersal rate, and relatively more dependent on the fundamental biodiversity number, as the connectivity or "panmixis" of the metacommunity decreases.*

Finally, figure 6.15 also gives an answer to the last remaining question of Williamson (number 3), namely, why do so many slopes of species-area relationships fall between 0.15 and 0.40? The unified theory's explanation is simply that a broad region of $θ - m$ parameter space exists for which z values lie between 0.15 and 0.4. Note that we are referring here to species-area relationships within continuous biogeographic landscapes (Rosenzweig's type 2 curves), not to the species-area curves obtained over an archipelago of discrete islands or habitat patches of different sizes (Rosenzweig's type 3 curves).

I now consider very large spatial scales, larger than the correlation length L, i.e., the spatial scale on which the species-area curves begin to steepen. At these large spatial scales, biogeographic processes become less and less correlated as the spatial sampling scale exceeds the distance separating dynamically independent biogeographic regions (Rosenzweig's type 4 curves).

It is possible to define a measure of the correlation length that is more informative of the actual spatial scale on which

189

the inflection point in the species-area curve occurs than the metric proposed by Durrett and Levin (1996). Recall that Durrett and Levin defined the correlation length L as the inverse square root of the speciation rate. They were interested in finding the upper bound of the slope of the species-area curve, and for this purpose it was useful to pick a very large number as a yardstick of large spatial scales. However, extensive simulation studies indicate that the upturn in the species-area curve always occurs well before a sampling area corresponding to L^2 is reached. In the metacommunity illustrated in figure 6.16, for example, the speciation rate is equal to $3 \cdot 10^{-6}$, which yields a Durrett-Levin corre-

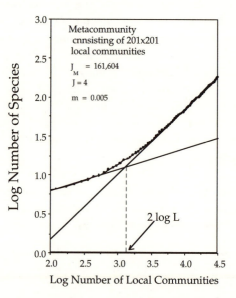

FIG. 6.16. The increase in the slope of the species-area relationship at large spatial scales, illustrated for the case of a metacommmunity of 201×201 local communities each of size $J = 4$, for $\theta = 1.0$ and $m = 0.005$. The inflection point defined by the intersecting tangents of the curve for intermediate and large areas yields an estimate of $2 \log L = 3.18$ local communities, which gives an estimated correlation length, L of 155.6 individuals. Mean of one hundred simulation runs.

lation length L of 568.5 individuals, or 142.1 for L in terms of local communities of size $J = 4$. The corresponding log area is $2 \log L$, or 4.31, which considerably over-estimates the observed inflection point (3.18) (fig. 6.16).

To find the inflection point, construct the tangent slopes to the species-area curve for intermediate (regional) and large spatial scales (fig. 6.16). The intersection of the tangent lines defines the inflection point in the species-area curve. In the present example, the estimated log area ($2 \log L$) at the inflection point is 3.18 local communities, which gives a value of $L = 155.6$ individuals, more than an order of magnitude smaller than Durrett and Levin's L. Hereafter, I will refer to the "correlation length L" as the midpoint in the upward inflection of the species-area curve, rather than the parameter of the same name defined by Durrett and Levin (1996). It should be noted that the slope of the upper tangent line of the species-area curve in figure 6.16 is not unity. However, a slope of unity is only theoretically expected for an infinitely large sample area. Observed species-area curves for finite biogeographic regions in nature will only rarely approach a slope of unity even at the largest spatial scales. In any event, this presents no theoretical or practical difficulty, because the tangents method for estimating L will yield the realized correlation length that actually characterizes a given biogeographic region. As I will discuss below, a geographical constraint to the free dispersal of species is expected to affect the realized correlation length, causing it to deviate from the predicted L based on θ, m, and unfettered metacommunity drift on an infinite homogeneous plain.

Throughout this chapter I have consistently used individuals as a surrogate for area. One can always convert individuals to an area measure according to our first principle (chapter 3). Thus, for the metacommunity illustrated in figure 6.16, the correlation length is approximately the distance across 36.3 local communities stretched end to end ($J = 4$).

Figures 6.17 and 6.18 illustrate how the species-area curves at large spatial scales and the correlation length L are controlled by the fundamental biodiversity number θ and by the dispersal parameter m. In figure 6.17 I have varied θ over two orders of magnitude from 0.1 to 10, holding m constant at 0.005. From figures 6.14 and 6.15, we would predict that increasing θ for a fixed value of m will raise the local species richness (α-diversity) as well as the slope (z value, β-diversity) of the species-area curve at intermediate spatial scales (figs. 6.17, 6.18). As θ increases and the species-area slope steepens, the inflection point in the curve becomes less pronounced because the difference in slope

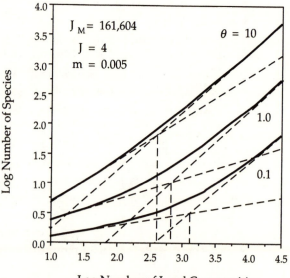

Fig. 6.17. Species-area relationships at large spatial scale for a metacommunity of 201×201 local communities, $J = 4$. Curves are drawn for $\theta = 0.1$, 1.0, and 10.0, for a low migration rate ($m = 0.005$). The correlation length L decreases with increasing θ for small m. Mean of one hundred simulation runs.

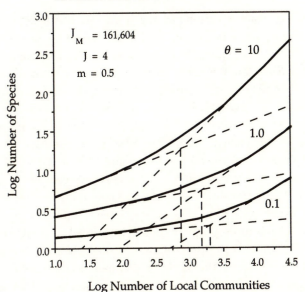

FIG. 6.18. Species-area relationships at large spatial scale for a meta-community of 201×201 local communities, $J = 4$. Curves are drawn for $\theta = 0.1$, 1.0, and 10.0, for a high migration rate ($m = 0.5$). The correlation length L increases with decreasing θ for large m. Mean of one hundred simulation runs.

above and below the inflection point becomes less and less. As θ increases, the upper and especially the lower tangent lines increase in slope. In figure 6.17, the upper tangent line increases from $z = 0.886$ for $\theta = 0.1$ to $z = 0.961$ for $\theta = 10.0$. The correlation length L *decreases* with increasing θ for a given value of m. This is because increasing the speciation rate increases the number of species in the meta-community, which reduces the mean size of populations and their geographic ranges because of the zero-sum rule. However, the correlation length *increases* with increasing dispersal rate m for a given value of θ. Increasing dispersal distributes common metacommunity species more widely, and couples the dynamics of local communities more broadly

193

geographically, enlarging average range sizes. Increasing the dispersal rate also reduces total metacommunity diversity, a phenomenon that will be examined more fully in chapter 7.

The correlation length L is a really important concept and number because it measures and defines the natural length scale of a biogeographic process over which metacommunity events are dynamically and evolutionarily connected. It is an especially important number for conservation biology because it quantifies the size and region within which observed metacommunity biodiversity evolves, lives, and dies. Thus, the unified theory assumes that, besides relative abundance distributions, species-area curves also contain information about both the fundamental biodiversity number and mean dispersal rates throughout the metacommunity. Although the slope of the species-area curve does not functionally determine the values θ and m, z does constrain θ and m to lie along a single isocline in θ-m parameter space (fig. 6.15). Conversely, knowing θ and m, according to the unified neutral theory, is sufficient to predict the slope of the species-area power law relationship expected on intermediate spatial scales.

The theory developed here for species-area curves makes the assumption that speciation rate is independent of dispersal rate. In general, one would expect speciation and migration rates to be inversely related (Mayr 1963, Slatkin 1977, 1980). High rates of dispersal should promote panmixis and therefore inhibit the speciation rate, at least for some modes of speciation. As we have seen, however, even assuming no interaction between dispersal and speciation, high rates of dispersal lead to lower slopes of species-area curves. A negative effect of dispersal on speciation would reduce the slopes of species-area curves still further. In the absence of quantitative data on the effect of dispersal on speciation rate, there is little beyond this qualitative statement that can be said here.

Before I conclude this chapter, it is useful to illustrate the application of the theory to interpreting species-area curves. My example consists of species-area relationships in twenty seven families of Panamanian trees and shrubs, compiled from regional checklists in the *Flora of Panamá* (D'Arcy 1987). Species-area curves were constructed for trees and shrubs achieving stem diameters > 1 cm dbh. The areas within which nested species counts were made ranged from the 50 ha plot (0.5 km^2) on Barro Colorado Island (BCI) at the low end, to all of BCI (15 km^2), to the area of the former Canal Zone (10^3 km^2), to the province of Panama ($2 \cdot 10^4$ km^2), and finally to the entire country of Panama ($7.5 \cdot 10^4$ km^2). From these data I computed the z values of the species-area curves for each family. Figure 6.19 presents the species-area curves for a representative sample of 6 families. Three of these are primarily families of midstory, canopy, or emergent trees in Panama (Bombacaceae, Myristicaceae, and Lecythidaceae), and three other families are generally shrubs or understory trees (Melastomataceae, Piperaceae, and Rubiaceae). I calculated the z values for the log-log linear portions of the curves, up to the area of the former Canal Zone.

Of the six sample families, the lowest z value occurred in the family Bombacaceae ($z = 0.097$), a family of mainly emergent, late secondary heliophilic species with small, wind- or bat-dispersed seeds (Croat 1978). Steeper curves were found in families of understory trees and shrubs (Melastomataceae, $z = 0.201$, Piperaceae, $z = 0.195$, and Rubiaceae, $z = 0.190$). Even steeper curves were found in the New World nutmeg family, Myristicaceae ($z = 0.215$) and in the brazilnut family, Lecythidaceae ($z = 0.262$). Myristicacs have large arillate, seeds dispersed by large frugivorous birds, but seldom far. Lecythidacs are a family of mainly midstory tree species, but with relatively poor seed dispersal. They have very heavy fruits containing large seeds that are eaten and scatter-hoarded by ground-foraging mammals.

195

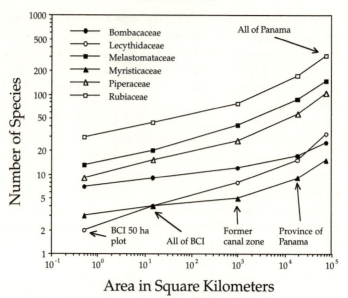

FIG. 6.19. Species-area curves for trees and shrubs > 1 cm dbh in 6 plant families in Panama, over a range of spatial scales from the 50 ha plot on BCI to all of Panama. Note the log-log linear curves at intermediate scales, up to the area of the former Canal Zone. It is suggestive and perhaps significant that all of the cures show a sudden upturn at approximately the same spatial scale, which corresponds to a correlation length of 70—130 km, the average width of the isthmus of Panama.

The diversity of species-area curves in figure 6.19 implies that the assumption of identical per capita dispersal and speciation probabilities among these families has been violated. However, as it turns out, the unified theory is quite robust to violations of this assumption, so long as the zero-sum rule still applies (chapter 10).

While it is perhaps not surprising that each of these plant families shows a unique species-area curve, what is striking is that all show a sudden upturn at about the same spatial scale (fig. 6.19). If we take this as a rough estimate of the correlation length, then it ranges from about 70 km to

130 km, approximately the average width of the isthmus of Panama. All but two of the twenty seven plant families show a similar inflection point. This strongly suggests that biogeographic processes of origination, dispersal, and extinction in the isthmus of Panama are constrained in some fashion by the fact that they are taking place on a long, narrow piece of land. This suggests that the geometry of a particular biogeographic area may override the theoretical expectations that each species-area curve should have its own intrinsic and endogenously generated correlation length.

We can estimate the approximate correlation lengths of the six Panamanian tree and shrub families in figure 6.19. All six curves are log-log linear up to a spatial scale of approximately 10^3 km^2, the area of the former Canal Zone. Then all begin to bend upwards, more or less at the same spatial scale, steepening to the area of the province of Panama, and steepening once again to the area of the entire country of Panama. Note that the curves, although becoming steeper, do not attain a slope of unity by the time all of Panama has been sampled.

I used the smallest three sample areas for each family to estimate the lower tangent line, and the two largest sample areas—the province of Panama and all of Panama—to estimate the upper tangent line. These yielded the following L^2 and correlation length L values for each of the six families: Bombacaceae: $L^2 = 11,000$ km^2, $L = 105$ km; Lecythidaceae: $L^2 = 16,000$ km^2, $L = 126$ km; Melastomataceae: $L^2 = 5,700$ km^2, $L = 75$ km; Myristicaceae: $L^2 = 7,000$ km^2, $L = 84$ km; Piperaceae: $L^2 = 6,800$ km^2, $L = 82$ km; and Rubiaceae: $L^2 = 4,600$ km^2, $L = 68$ km. It is interesting that all of the estimated correlation lengths are small and relatively similar, varying only over about a 2 fold range, from 68 km to 126 km. Also, the three plant families that are primarily understory shrubs and treelets (Melastomataceae, Piperaceae, and Rubiaceae) had the smallest correlation lengths.

The historical, ecological, and evolutionary explanations for the variation in species-area curves among Panamanian tree and shrub families are currently unknown. Speciation rate aside, however, one might expect that dispersal rates would be affected by plant growth form. Canopy trees might be expected to disperse seeds on average farther and faster than small-stature understory shrubs or herbs. Therefore, plant families that are comprised mainly of canopy or emergent tree species might be expected to have shallower species-area curves than families comprised mainly of shrubs or herbs. The species-area data for the twenty seven Panamanian tree and shrub families show generally good agreement with this prediction (fig. 6.20). On intermediate spatial scales, families comprised mainly or exclusively of canopy or emergent tree species exhibited species-area curves having consistently lower z values than families comprised mainly of understory trees, treelets, or shrubs.

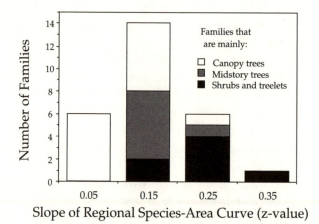

Fig. 6.20. Relationship between the characteristic stature of species in 27 tropical tree and shrub families in the flora of Panama and the slope (z-value) of the Arrhenius log-log linear species-area curves for families at intermediate spatial scales in Panama. Families comprised mainly of small-stature species have steeper species-area curves than families of large-stature species.

When we examine the species-area curve for the complete native tree and shrub flora of Panama, we see a similar upturn in the species-area curve at about the same spatial scale (fig. 6.21). Drawing the upper and lower tangents to the species-area curve yields an L^2 of 4300 km^2, and a correlation length of 65.6 km, which once again is commensurate with the average width of the Panamanian isthmus. We can also estimate the correlation length for the vascular plant flora of the world from fig. 6.3 (Shmida and Wilson 1985). The value of the squared correlation length is approximately $7 \cdot 10^{11}$ m^2, yielding an estimated correlation length L of $8.4 \cdot 10^5$ m, or 837 km. For the world's avifauna,

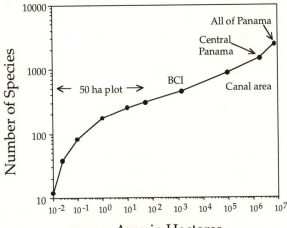

Area in Hectares

FIG. 6.21. Species-area curve for the entire native tree and shrub of Panama, over a range of spatial scales from 10^{-2} ha (a 10× 10m quadrat within the 50 ha BCI plot) to all of Panama. In Panama, the range of intermediate spatial scales over which the log-log linear Arrhenius relationship applies is approximately from 1 ha to 10^5 ha, suggesting that the mean correlation length for the Panamanian tree flora lies somewhere in the range of 30—300 km. Data extracted from D'Arcy (1987). Note similarity in shape to Shmida and Wilson's (1985) species-area curve for the world's flowering plants (fig. 6.3).

the squared correlation length is approximately $2.5 \cdot 10^6$ km^2 (fig. 6.2), yielding a correlation length L of 1585 km.

SUMMARY

1. When species-area curves are plotted from local to global scales, they are triphasic, indicating that different scaling rules apply on different spatial scales. As a result, the Arrhenius relationship, a linear log-log species-area power law—which implies self-similarity in the intermediate scaling region for which it applies—does not work at small and very large spatial scales.

2. The unified theory asserts that these triphasic species-area curves are the spatial manifestation of the steady state of a dynamical, neutral speciation-dispersal-extinction process playing out on the metacommunity landscape. The theory explains the triphasic nature of species-area curves without resorting to niche-assembly hypotheses.

3. On local spatial scales, species-area curves are sensitive to the local relative abundance of species and are curvilinear on a log-log plot.

4. On regional spatial scales, the theory always predict's linear log-log species area curves whose slopes (z values) depend on the fundamental biodiversity number θ and m, i.e., on the relative rates of speciation and dispersal across the metacommunity landscape.

5. At large spatial scales, the neutral theory predicts the existence of a natural length scale, the correlation length L, which defines the mean size of evolutionary-biogeographic units on the metacommunity landscape; L is a function of θ and m.

6. As spatial sampling scales exceed L, the slope of the log-log species-area curve steepens toward unity as evolutionary-biogeographic processes on regional

scales become increasingly uncoupled and dynamically uncorrelated with one another.

7. The spatially explicit theory needed for species-area relationships differs from the spatially implicit theory for the classical island-mainland problem posed by MacArthur and Wilson in that *both* the fundamental biodiversity number θ and the dispersal rate m must be specified.

CHAPTER SEVEN

Metapopulations and Biodiversity on the Metacommunity Landscape

The theory of island biogeography has been criticized in recent years for failing to take into account the fragmented nature of populations and the habitats that they occupy (e.g., Hanski and Simberloff 1997). Perhaps this criticism has some validity when applied to the classical island-mainland problem posed in the original theory. But in fairness, MacArthur and Wilson did consider the more complex problem of the biogeography of archipelagos of islands or habitats. Since their monograph appeared more than 30 years ago, great strides have been made in understanding and mathematically characterizing complex landscapes (Mandelbrot 1982, Milne 1997, Ritchie 1997, Ritchie and Olff 1999). Indeed, the whole field of fractal geometry did not exist then (Gleick 1987); but there is no reason in principle why the theory cannot encompass fractal landscapes and metapopulation biology. Probably the most important reason for incorporating the metapopulation perspective is the practice of conservation biology. Although natural habitats have always been patchy, the anthropogenic destruction of natural habitats has greatly worsened the problem of habitat fragmentation (Tilman et al. 1994), and fragmentation is a fact of life that is here to stay.

How are the conclusions we have reached in previous chapters affected by habitat fragmentation? How does the metacommunity equilibrium distribution of relative species abundance change on a fragmented landscape? How are the expected times to extinction of individual species affected by habitat fragmentation under the neutral theory? A full examination of these questions is beyond the scope of present work, and lies in the future. However, there are a number of initial questions we can explore with the existing theory. I first consider a single-species population and ask: Under ecological drift and random dispersal, what is the probability that a species is present or absent from a local community, as a function of local community size, the probability of immigration, and the metacommunity relative abundance of the species? Then I consider an archipelago of islands or habitats and ask: What is the covariance of abundance of the ith species among islands or habitat patches? Finally, I consider metacommunities and ask: How is biodiversity explicitly spatially distributed on a continuous metacommunity landscape?

One of the most important questions for the conservation of particular focal species is the probability that the species is present in a given patch. This is formally equivalent to the proportion of time that the species occupies the habitat patch. We can study such incidence functions from the theory developed in chapter 4 for the dynamics of single species undergoing zero-sum ecological drift. The *incidence function* specifies the equilibrium fraction of time that the ith species will be present in the local community, which is given by the sum of the elements of the eigenvector $\Psi(n)$ for $N_i \geq 1$ (see chapter 4). Increasing both m and P_i increases the incidence of the ith species in the local community (fig. 7.1). An important conclusion from figure 7.1 is that it does not take a very high rate of immigration or a high metacommunity relative abundance of the focal species to have a high probability of being present in a local community. However, obviously the immigration rate must be nonzero, and there must be a metacommunity from which immigrants can come. This result is from the implicit

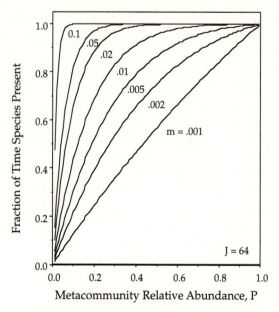

FIG. 7.1. Equilibrium incidence functions for the ith species in an ergodic community undergoing zero-sum drift, as a function of probability of immigration m, and metacommunity relative abundance P_i for a local community of size $J = 64$.

space version of the theory (see Chapters 4 and 5), and it assumes a very stable metacommunity from which immigrants are drawn, stabilized by the law of large numbers.

Increasing the size of patches also has a large effect on the probability that a species will be present in the given patch or local community (fig. 7.2). For example, if the patch or local community size J is 10^5, then an immigration rate of 10^{-3} will maintain the presence of a rare metacommunity species (1% of the metacommunity) essentially 100% of the time. Larger communities also lower the extinction rate. This is because larger communities allow larger population sizes to develop, which in turn delays the inevitable local extinction. This is the primary insight that led MacArthur

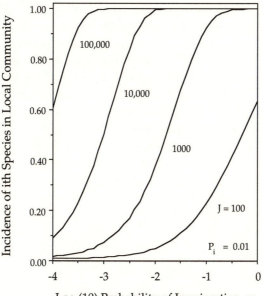

FIG. 7.2. Equilibrium incidence functions for the ith species in an ergodic community undergoing zero-sum drift, as a function of probability of immigration m, and a metacommunity relative abundance $P_i = 0.01$, for four orders of magnitude variation in local community of size.

and Wilson (1967) to draw extinction rate curves as a function of island size in their graphical model of island biogeography. The smaller the community size or habitat patch of individuals, the more likely a species is to be absent and the more variable the species composition of the community will be. Also, the more isolated a community is, the more likely a given species is to be absent, in accordance with MacArthur and Wilson's theory.

Another perspective is obtained from calculating the persistence function. The *persistence function* gives the mean passage time of a species between abundance i and abundance j. A particularly important persistence function is that which characterizes the mean passage time between colonization and extinction events of a species in the local community or island. The shortest passage times to extinction will be found in species with lowest abundances, which will include most recent colonists. However, if a species manages to become common in the community, its expected passage time to extinction will increase. The mean persistence times can be calculated for the ergodic community as follows. For any ergodic process, which is defined as one in which every state is ultimately reachable from every other state, there exists a matrix $T = \{t_{ij}\}$ of the mean number of steps to reach abundance j for the first time, starting in abundance i. Matrix T is given by

$$T = W - W_{dg}$$
$$W = \{I - E + CE_{dg}\}D,$$

where I is the identity matrix, E is the fundamental matrix of the ergodic process, C is a matrix whose entries are all unity, W_{dg} and E_{dg} are matrices with the same elements on the principal diagonal as matrices W and E, respectively, and zeros elsewhere, and D is a diagonal matrix with each element on the principal diagonal equal to $1/\Psi_j$. The fundamental matrix E is given by

$$E = \{I - B + \Psi\}^{-1},$$

where B is the matrix of transition probabilities for the ergodic process, and Ψ is a matrix whose elements are ψ_i across each row (see chapter 4).

It is instructive to consider again our simplest ergodic community, when $J = 1$. The mean passage time to extinction for the resident species is easily found from the above equations to be

$$T_{N_i=1 \rightarrow N_i=0} = \frac{1}{m(1 - P_i)}.$$

This demonstrates that persistence time in the patch or local community is inversely related to the immigration rate m and also inversely related to the collective relative abundance of all species other than species i in the metacommunity. As the ith species increases in metacommunity relative abundance toward monodominance (i.e., as $P_i \to 1$), the persistence of the ith species tends toward permanent occupancy ($T_{N_i=1} \to \infty$) of any given local site. Also note that the site is permanently occupied by species i (i.e., by whatever species is initially present) if there is no immigration ($m = 0$). The inverse relationship between persistence and immigration rate may seem counterintuitive. However, recall that the immigration rate in MacArthur and Wilson's original model applied to all species—not just a given focal species. Therefore, *increasing the immigration rate has the effect of increasing the rate of species turnover in a patch or local community*, displacing resident species by other species immigrating into the patch from the metacommunity.

The mean passage time to extinction is the shortest for $N_i = 1$. For initial abundances $N_i > 1$, the mean passage times become longer and the persistence functions become more complex in communities of size $J > 1$. Figure 7.3 shows the case of a patch or local community of size $J = 16$, and the persistence functions for $N_i = 1$, $J/2$, and J. The persistence functions are now U-shaped for $J > 1$. At very low and very high rates of immigration, the persistence times are longer than at intermediate rates of immigration. At low rates of immigration, there is no force of immigration to increase the turnover rate of species in the patch. At very high rates of species immigration, the immigration provides a subsidy of the species in the patch, and this subsidy delays the local extinction of the ith species. This is the unified theory's formal confirmation of the "rescue effect" postulated by Brown and Kodric Brown (1977). The parallels with the absorbing case discussed in chapter 4 are obvious.

207

Among the elements of Var(T) are the variances in passage times to extinction from any starting abundance. As in the absorbing case (chapter 4), the time to local extinction in the ergodic community is approximately gamma distributed.

I now consider the covariance of the abundance of the ith species between two discrete habitat patches, islands, or local communities. Imagine an archipelago of habitat patches or islands and let this archipelago be bathed in a background immigration rate. Now consider two islands, and imagine that they are dynamically coupled both through the background metacommunity immigration, and through the exchange of migrants specifically between the two islands or habitat patches. The expected total abundance of the ith species in the two equal-sized habitat patches or islands of size J is simply $2JP_i$. But how does the abundance of the ith species covary in these habitat patches as a function of reciprocal migration? Presumably, local communities that are immediately adjacent to one another will be more similar in abundance of the ith species than communities separated by greater distance. These local communities are dynamically coupled to one another, and the strength of this coupling depends on the rate of exchange of migrants, the size of the local communities, as well as the fundamental biodiversity number. This covariance in the ith species among two local communities can be studied analytically.

Consider the dynamics of the ith species in the context of the ergodic community studied in chapter 4. Now, however, let us consider the coupled dynamics of two local communities that exchange migrants per birth with probability m', and let m once again be the probability that either local community receives an immigrant from the metacommunity in which they are imbedded. Note that m' may be larger than m, but it will often be smaller than m because the metacommunity typically is a much larger source area than a single other local community or another island in an archipelago, and so it has a higher "mass effect" (Shmida and Ellner 1984). Let N_i and O_i be the abundances of the ith species in the two local communities, respectively, each of which is of size J. Assume $D = 1$ for simplicity. We can now write down the transition probabilities for changes in the joint abundance of the ith species:

$$\Pr\{N_i - 1, O_i | N_i, O_i\}$$
$$= \frac{N_j}{2J}\Big[m(1 - P_i) + (1 - m)m'\Big(\frac{J - O_i}{J}\Big)$$
$$+ (1 - m)(1 - m')\Big(\frac{J - N_i}{J - 1}\Big)\Big]$$

$$\Pr\{N_i, O_i - 1 | N_i, O_i\}$$
$$= \frac{O_i}{2J}\Big[m(1 - P_i) + (1 - m)m'\Big(\frac{J - N_i}{J}\Big)$$
$$+ (1 - m)(1 - m')\Big(\frac{J - O_i}{J - 1}\Big)\Big]$$

$$\Pr\{N_i, O_i | N_i, O_i\}$$
$$= \frac{N_j}{2J}\Big[mP_i = (1 - m)m'\Big(\frac{O_i}{J}\Big)$$
$$+ (1 - m)(1 - m')\Big(\frac{N_i - 1}{J - 1}\Big)\Big] + \frac{J - N_i}{2J}$$
$$\times \Big[m(1 - P_i) + (1 - m)m'\Big(\frac{J - O_i}{J}\Big)$$
$$+ (1 - m)(1 - m')\Big(\frac{J - N_i - 1}{J - 1}\Big)\Big] + \frac{O_i}{2J}$$
$$\times \Big[mP_i + (1 - m)m'\Big(\frac{N_i}{J}\Big)$$
$$+ (1 - m)(1 - m')\Big(\frac{O_i - 1}{J - 1}\Big)\Big] + \frac{J - O_i}{2J}$$
$$\times \Big[m(1 - P_i) + (1 - m)m'\Big(\frac{J - N_i}{J}\Big)$$
$$+ (1 - m)(1 - m')\Big(\frac{J - O_i - 1}{J - 1}\Big)\Big]$$

$$\Pr\{N_i + 1, O_i | N_i, O_i\} =$$
$$\frac{J - N_i}{2J}\Big[mP_i + (1 - m)m'\Big(\tfrac{O_i}{J}\Big)$$
$$+ (1 - m)(1 - m')\Big(\tfrac{N_i}{J-1}\Big)\Big]$$

$$\Pr\{N_i, O_i + 1 | N_i, O_i\} =$$
$$\frac{J - O_i}{2J}\Big[mP_i + (1 - m)m'\Big(\tfrac{N_i}{J}\Big)$$
$$+ (1 - m)(1 - m')\Big(\tfrac{O_i}{J-1}\Big)\Big].$$

Note that these equations differ from those for the single ergodic community in having an additional term for immigration from the other local community. Also note that each probability is multiplied by one-half. This is because of the condition that $D = 1$ and the single death per death-birth cycle occurs in one or the other local community with equal probability, 0.5.

We can once again compute the eigenvector for this system of equations. For the simplest case of two local communities of size $J = 1$, the eigenvector is

$$\Psi_1(N_i, O_i) = \begin{pmatrix} N_i = 0, O_i = 0 \\ N_i = 1, O_i = 0 \\ N_i = 0, O_i = 1 \\ N_i = 1, O_i = 1 \end{pmatrix}$$

$$= \begin{pmatrix} (1 - P_i)(1 - mP_i) \\ (1 - P_i)mP_i \\ (1 - P_i)mP_i \\ P_i[1 - m(1 - P_i)] \end{pmatrix}.$$

When two local communities are considered jointly, the $J = 1$ case is no longer independent of the probability of immigration from the metacommunity, m, as it was in chapter 4. However, the $J = 1$ case *is* independent of the probability of migrants from one local community to the other m'. Note that as $m \rightarrow 1$, $\Pr\{0,0\} \rightarrow (1 - P_i)^2$ and $\Pr\{1,1\} \rightarrow P_i^2$. As $P_i \rightarrow 1$, $\Pr\{0,0\} \rightarrow 0$ and $\Pr\{1,1\} \rightarrow 1$; conversely, as metacommunity abundance $P_i \rightarrow 0$, $\Pr\{0,0\} \rightarrow 1$ and $\Pr\{1,1\} \rightarrow 0$.

The eigenvector for arbitrary $J > 1$ is algebraically extremely messy, but it is easy to compute exactly. From this eigenvector, we can directly calculate the covariance of the abundance of the ith species in the two local communities. The expected abundance in each local community is identical to the expectation derived in chapter 4 in the single local community case: $E\{N_i\} = E\{O_i\} = JP_i$. Therefore, the covariance in abundance of the ith species is

$$\mathrm{Cov}\{N_i, O_i\} = \sum_{N_i=1}^{\prime} \sum_{O_i=1}^{\prime} (N_i - JP_i)(O_i - PJ_i)\Psi_j(N_i, O_i).$$

Figure 7.4 illustrates the functional dependence of the covariance in the abundance of the ith species on the probabilities of immigration from the source metacommunity m, and from the other local community m', for a local community size $J = 8$. The covariance exhibits contrasting dependence on m vs. m'. Increasing the migration rate from the metacommunity *decreases* local community covariance, whereas increasing the reciprocal migration rate between the local communities *increases* their covariance. The covariance between local communities is zero when $m = 1$ and $m' = 0$. Conversely, the local communities maximally covary when $m \rightarrow 0$ and $m' = 1$, in which case the limiting maximal covariance is $(JP_i)^2$.

We can now answer the question of how the abundances of the ith species in two local communities will covary as a function of their dynamic coupling through reciprocal migration. We can measure distance between local communities of equal size J in units of $\sqrt{J/\rho}$. In the last chapter

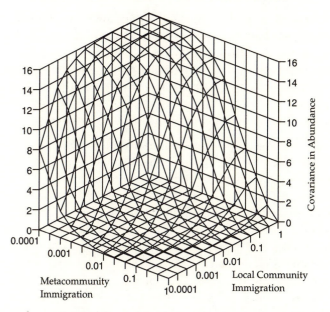

FIG. 7.4. Covariance of the abundance of the ith species in two habitat patches or islands as a function of the probability of immigration from the metacommunity, m, and from the other local community, m'. In this example, local community size $J = 8$ and metacommunity abundance $P_i = 0.5$.

we noted that, on local spatial scales, the probability of immigration to a local community from the metacommunity will fall off approximately as some power function of local community size, $m(J) = J^{-\omega}$. Assuming that the background immigration rate m from the metacommunity is not affected by varying the spatial separation of the two local communities, and because J increases linearly with area, then we simply compute the covariance as a function of $m' = n(J/\rho)^{-\omega'/2}$, where ω' is empirically measured, and $n = 1, 2, 3, \ldots$.

The preceding analysis is most appropriate for a fragmented landscape or an archipelago of islands where one can treat space implicitly. I have not worked out the covariance when the two local communities are imbedded in a landscape in an explicitly spatial model. In this case, there will be distributed effects of dispersal from near and far communities, and the covariance function will be considerably more complex than the case computed here. This is another example of the many interesting theoretical problems in the neutral theory that remain for the future.

I turn now to a consideration of the distribution of total biodiversity on the metacommunity landscape. Although there are few analytical results for the explicit spatial case, the results of many simulations lead to a number of important qualitative generalizations that are consistent with the analytical results presented in chapters 4 through 6. As we have seen in chapter 6, the steady-state species-area relationship is controlled by the fundamental biodiversity number θ and by the probability of dispersal m. This implies that speciation and dispersal limitation ($m < 1$) will affect the distribution and maintenance of biodiversity on local to regional scales under zero-sum ecological drift, and the spatial autocorrelation of community species composition. We can illustrate several of the most important generalizations with the results of a numerical experiment, simulating a metacommunity consisting of 101×101 local communities, each of size $J = 16$. I initialized diversity at one individual per species, and ran simulations for 100,000 birth-death cycles with four deaths per local community per cycle, resulting in 25,000 complete turnovers of the metacommunity. In this example, the fundamental biodiversity number θ was set at 14, a reasonable value, say, for a southern temperate forest. Let dispersal in one time step be restricted to the Moore neighborhood (chapter 6) of any given local community.

Now consider two cases: case I, in which migration rates are very low ($m = 0.005$) and local communities are quite isolated from one another; and case II in which migration rates between neighboring communities are very high

($m = 0.5$) and local communities are quite connected by dispersal. In case II, one out of every two births is an immigrant, whereas in case I, immigration accounts for only one out of every two hundred births.

What do we predict about the spatial structure of species populations in the metacommunity from the neutral theory? Recall from the analytical results of chapter 4 that under severe dispersal limitation, a focal species is expected to spend most of the time either locally extinct or, less often, monodominant. This implies that local communities will become less diverse, lose species, and show increased dominance under low immigration rates and greater isolation. If dispersal rates are low, then some species or other gets carried by ecological drift to high local abundance and potentially monodominance. Conversely, if the dispersal rates are high, then local communities will have more species and monodominance will be rare or absent. In figure 7.5 I show sample results for case I (low dispersal rate), and I have plotted the distributions of four representative species in a block of 21×21 local communities in the center of a 101×101 metacommunity: two common species, a species of intermediate abundance, and a rare species.

Under case I, when dispersal is extremely limited, a tile-like mosaic pattern becomes apparent (fig. 7.5). Common and rare species alike show a strong tendency toward monodominance in their respective patches. Thus, low rates of migration strongly reduce local community species richness and increase local dominance. Note that these monodominant local communities tend to have occasional satellite individuals in adjacent communities, but rarely more than one or two individuals at a time. Thus, as predicted, strong dispersal limitation reduces local community species richness and increases local community dominance. This effect becomes stronger as $m \rightarrow 0$. The spatial distributions of species under case II are completely different from case I (fig. 7.6). When local communities are strongly

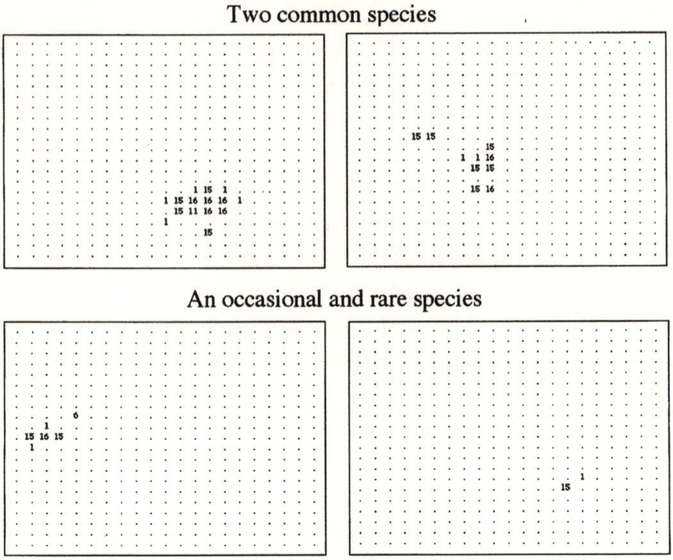

Two common species

An occasional and rare species

FIG. 7.5. case I: Local communities are very isolated ($m = 0.005$). Typical species distribution maps at steady-state under spatially explicit ecological drift in the central block of 21×21 local communities each of size $J = 16$. The numbers in the grids are the local abundances of the given species. Fundamental biodiverity number in the grids are the local abundances of the given species. Fundamental biodiversity number $\theta = 14$. The species in the top two panels are maps of two common species; the bottom two panels are an occasional (*left*) and a rare species (*right*), respectively. Note that all species form patches of partial to complete local monodominance producing a mosaic metacommunity. Total metacommunity size is 101×101 local communities.

coupled dynamically through high rates of dispersal, the tile-like mosaic of monodominant species completely disappears. In its place, we have much more amorphous and diffuse spatial distributions, and local community species richness is higher while dominance is lower. Because this is a stochastic equilibrium, the individual species distributions on the metacommunity landscape are not static but continue to move about as species come and go. Note also

216

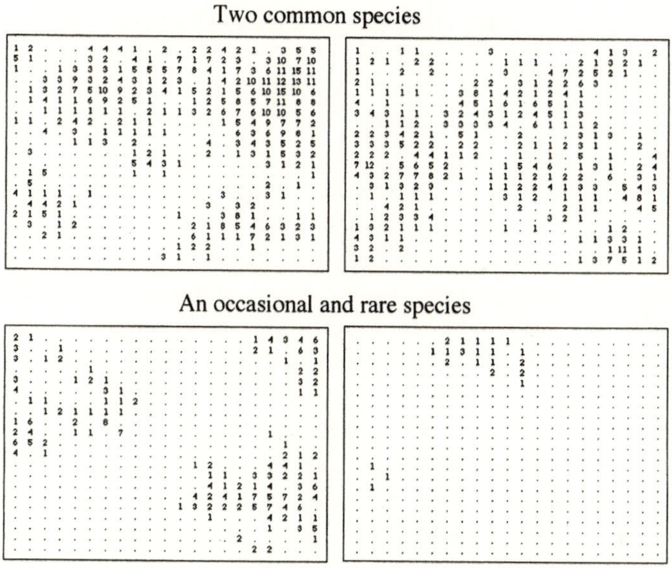

Two common species

An occasional and rare species

FIG. 7.6. case II: Local communities are highly coupled by migration ($m = 0.5$). Typical species distribution maps at steady-state under spatially explicit ecological drift in the central block of 21×21 local communities each of size $J = 16$. Numbers in the grids are the local abundances of the given species. Fundamental biodiversity number $\theta = 14$. The species in the top two panels are maps of two common species; the bottom two panels are an occasional (*left*) and a rare species (*right*), respectively. Note that species are much more diffuse in dispersion pattern, and much more intermingled in local communities than in case I. Total metacommunity size is 101×101 local communities.

that metacommunity populations of individual species may become fragmented into isolated demes, and look and behave like metapopulations. It is interesting that such fragmentation arises more frequently when dispersal rates are high rather than low.

These simulated spatial patterns are only caricatures of the much more complex patterns that would be found in actual natural communities, but they are nevertheless sufficient to make the following very important general

217

point. Under low rates of dispersal, species are found at higher local abundance in patches of lower local species richness. Conversely, under high rates of dispersal, species occur at lower abundance in local communities having higher species richness.

This means that there is a change in the distribution of α (local) and β (regional) diversity with a change in the dispersal rate (fig. 7.7). High rates of dispersal bring more regional diversity to the local community, but a consequence is that abundant and widespread metacommunity species drive rare and local species extinct from the metacommunity. The result is that total metacommunity biodiversity is reduced by high dispersal rates. Conversely, low rates of dispersal let many rare local "endemics" survive in small pockets of high abundance, so that total metacommunity biodiversity is increased. Low dispersal rates mean that ecological drift has adequate time between dispersal events or carry a species to high local abundance. Because dispersal is infrequent, there is little force of immigration to break up these local pockets of high abundance or monodominance. As was shown in chapter 4, monodominant species in local communities take much longer to drift to local extinction than less abundant species in the local community. Thus, the general result is that *under high rates of dispersal, local diversity is high but metacommunity diversity is low; whereas under low rates of dispersal, local diversity is low but metacommunity diversity is high.*

Note in figure 7.7 that the dominance-diversity curve for the metacommunity in the dispersal-limited case has somewhat of a staircase-like appearance. This is due to the high frequency of species having abundances that are multiples of the local community size J, which in this case was set to 16. Thus, there are greater numbers of species with abundances of 16, 32, 48, etc., than would otherwise be expected without such frequent local monodominance. This staircase is in essence a modeling artifact of having to simulate very

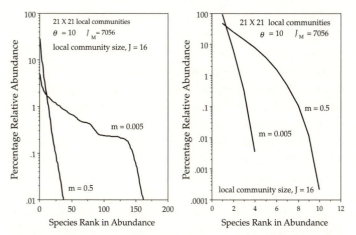

Fig. 7.7. Comparison of the steady-state dominance-diversity curves that arise under strong dispersal limitation (case I: $m = 0.005$), and under strong dispersal coupling (case II, $m = 0.5$) among local communities of size $J = 16$. *Left:* Metacommunity dominance-diversity. *Right:* Local community dominance diversity. Dominance-diversity curves were computed for the central block of 21×21 local communities in a metacommunity of 101×101 local communities after 100,000 birth-death cycles, and four deaths in each local community per cycle.

small discrete community sizes in the computer. With continuous intergradations of local communities, the frequency of perfect local monodominance becomes very small, and this staircase appearance disappears. One would not expect to see this effect in spatially continuous ecological communities in nature.

Let us now consider the question of how the similarity of communities changes with distance across the metacommunity landscape. Many ecological and evolutionary processes will produce a loss in similarity between two ecological communities with increasing separation distance. Under niche-assembly theory, distance decay is predicted to result from species turnover along local and regional environmental gradients or among habitats. In nature, habitats are often patchy and recurrent, so that the decay of community similarity with distance may frequently not be smooth. However, the neutral theory also predicts a decay in community similarity with distance, and it does so on completely homogeneous landscapes. Indeed, perhaps the smoothest decay in similarity with distance should be predicted by neutrality because no other forces are operating besides ecological drift, random dispersal, and random speciation. The reason for the decay in similarity is that large steady-state differences exist in the metacommunity abundances of species, coupled with dispersal limitation that is increasingly severe in ever rarer species. The steady-state distribution of metacommunity species abundance and the level of dispersal limitation are dictated by the fundamental biodiversity number θ and by the dispersal rate m. We have already seen that the neutral theory predicts species-area relationships chapter 6), so it should come as no surprise that the theory also predicts the distance decay in similarity in communities, which in a sense is simply the inverse problem.

In an excellent analysis of the distance decay of similarity in biogeography and ecology, Nekola and White (1999) analyzed data on the decay of similarity in several plant metacommunities. Perhaps the most relevant of their datasets for testing the neutral theory (because of its relative homogeneity) is one on boreal upland white spruce forests, a dataset collected by LaRoi (1967) and LaRoi and Stringer (1976). The dataset consisted of species lists in thirty-four 9 ha plots over a 6000 km transcontinental transect from Newfoundland to Alaska. These plots collectively contained 561 species of vascular plants. To measure similarity in communities, Nekola and White used Jaccard's Index (Mueller-Dombois and Ellenberg 1974), which is probably the most widely used and familiar index of similarity. They used a Mantel's test with 10,000 replications to obtain bootstrap estimates of the significance of their statistical models (Manly 1991).

With this and other datasets, Nekola and White (1999) were able to ask a number of fundamental questions, including the following. First, what is the functional form of the distance-decay curve? Second, does this curve depend on what component of the plant community is considered, such as trees, shrubs, or herbs? Third, do species of differing metacommunity abundance contribute differentially to the decay curve? Fourth, is there evidence of involvement of mode of dispersal in the rate of distance decay, i.e., do better dispersing taxa have slower rates of distance decay?

The similarity decay curves for trees, small shrubs, and herbs in upland white spruce forests across Canada and Alaska are shown in figure 7.8 over the 6000 km range. In answer to question 1, Nekola and White (1999) found that these decay curves were all fit best by simple negative exponential functions. In answer to question 2, the steepness of these decay curves was a function of plant growth form. It was shallowest for trees, and became progressively steeper the smaller the plant growth form (fig. 7.8). In answer to question 3, the component of abundant and widespread species had shallower decay curves than the component consisting of species of intermediate abundance (not illustrated). However, the rare-species component also showed very little decay in similarity because it was already maximally dissimilar due to the generally very local distribution of rare species. Finally, in answer to question 4, mode of dispersal did influence the decay of similarity. Animal-dispersed, fleshy fruited species had slower rates of decay than wind- or spore-dispersed species. These results are entirely congruent with the results presented in chapter 6 on the species-area relationships for the tree flora of Panama.

The neutral theory's qualitative predictions are completely consistent with Nekola and White's results, but they differ slightly in the predicted functional form of the decay curve

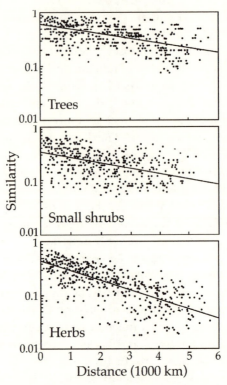

F$_{IG}$. 7.8. Distance decay in Jaccard's Index of Similarity for three components of upland white spruce forest from Newfoundland to Alaska over nearly 6000 km. Three components of the vascular plant community are sown: trees (*top*), small shrubs (*middle*) and herbs (*bottom*). Slopes of semilog plots are as follows: trees, −0.19; small shrubs, −0.28; and herbs, −0.40 After Nekola and White (1999).

because the expected curves are not perfect exponentials. And they should not be for the very reasons given by Nekola and White (1999): namely, that communities consist of mixtures of widespread and abundant metacommunity species, and of rarer and more local species whose distance decay is steeper. This heterogeneity will produce somewhat rounded distance-decay curves that are *compound exponential*, with a different exponent for each added species abundance class of the community. Typical curves that one obtains from the neutral theory are shown in an arithmetic plot in figure 7.9, and in a semilog plot in figure 7.10. They start more steeply because they are dominated by the turnover of rare and occasional species that are not widespread. However, as more and more of the local species drop out, the species that remain are the widespread and abundant species, so the distance-decay curve will begin to slow down and have a shallower and shallower slope.

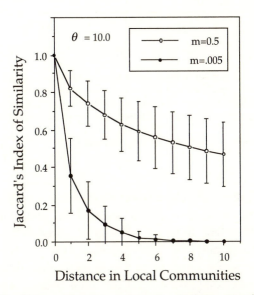

FIG. 7.9. Distance decay in community similarity according to Jaccard's Index. Decay in similarity is much faster for low rates than for high rates of dispersal. These curves were averages over one hundred simulations of a 101 × 101 metacommunity, with a local community size of $J = 8$ and a fundamental biodiversity number θ of 10. The curves are drawn for the central 21 × 21 community region in the middle of the metacommunity. Error bars are one standard deviation of the mean. These curves appear exponential on an arithmetic plot, but they are actually compound exponential. See figure 7.10.

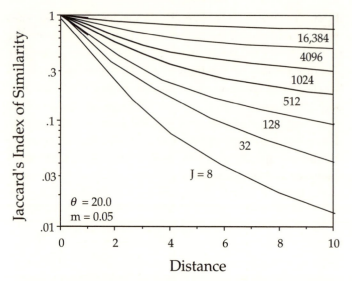

FIG. 7.10. Semilog plot of the distance decay of Jaccard's Index of Similarity, revealing the compound exponential nature of the decay curves. The curvilinearity is easier to detect when using a small sampling "grain size" or local community (i.e., small J) rather than large J, and when the decay curves are plotted semilogarithmically. Distance is measured in number of local communations.

Note that the distance-decay curve is strongly influenced by the mean dispersal rate over the metacommunity (fig. 7.9). High dispersal rates prevent differentiation of local communities through ecological drift. Conversely, low dispersal rates allow much more local differentiation of community composition, and the distance-decay curves are much steeper. Once again, this is completely consistent with the analytical results of chapter 4.

The curvilinearity of the distance decay is more obvious for small J and when the decay curves are plotted semilogarithmically. The degree of curvilinearity will depend on the degree of differentiation of relative species abundance in the metacommunity, which depends on the fundamental biodiversity number θ and on the dispersal rate m. This effect is seen in figure 7.10, in which I have varied J, the size of the local community as the sampling unit. The compound nature of the exponential distance decay curve will be harder to detect if the local community size is large. This corresponds to choosing a large spatial sampling unit or "grain size" *sensu* Palmer and White (1994). This effect happens because most of distance decay in similarity of the rare and local species will occur rapidly at shorter distances. Given the scatter in the data on distance-decay curves such as those shown in figure 7.9, and given the relatively slight curvature expected on these spatial scales, it may be difficult to detect the predicted curvilinearity in the distance-decay relationship. However, the very fact that shallower slopes were obtained for widespread taxa than for taxa of intermediate frequency indicates that this curvilinearity must be present in the datasets used by Nekola and white (1999).

The effects of dispersal on α and β diversity have major implications for conservation strategies. Changes in the mean dispersal rate cause changes in the spatial distribution of total biodiversity and the degree of local endemism across the metacommunity landscape. In recent papers, Harte and Kinzig (1997) and Harte et al. (1999) point out that the species-area curve for endemics has the same functional form as total species-area curves on intermediate spatial scales, i.e., $S = cA^z$, but the exponent z is much steeper for endemics than for total species richness. This phenomenon is predicted by the unified theory because a low dispersal rate has the effect of making species locally more abundant but also rarer and endemic to smaller regions of the metacommunity landscape as a whole. Thus, species-area curves are significantly steeper when dispersal limitation is greater, consistent with our results from the preceding chapter.

These results also have important implications for the twenty-five-year-old debate about reserve design, and the wisdom of having one large reserve versus many small scattered reserves (Diamond and May 1975, Wright and Hubbell 1983, Burkey 1995). From the perspective of the unified theory, this question cannot be resolved definitively without understanding the degree of dispersal limitation affecting the metacommunity and how dispersal limitation will be affected by fragmentation of the landscape. The message from the theory, therefore, is that we need to gather much better and more extensive information about dispersal ability and actual movements of species over the metacommunity landscape. Qualitatively we can say, however, that the more dispersal limited or fragmented the metacommunity, the more likely that a multiple reserve design will be necessary to save local endemics. This point was also stressed by Harte and Kinzig (1997).

Tilman et al. (1997) discussed the trade-off between dispersal limitation and competitive ability in this context. They argued that good competitors, which are species capable of holding onto sites but are generally not good dispersers, will suffer the most from fragmentation of the landscape because their movement will be more affected by fragmentation than the movement of good dispersers. Species in the unified theory do not exhibit this trade-off, and so the present neutral theory does not make such a prediction. However, it does predict that species that are common and widespread throughout the metacommunity prior to fragmentation will also be the most resistant to the effects of fragmentation. Thus, the unified theory asserts that the most critical factor determining the long-term survival of a species is its metacommunity abundance and distribution. We have seen that widespread species are more "competitive" in the sense that they persist longer than rare species, and they can even hasten the demise of rare species if dispersal rates are increased. In this sense, fragmentation may actually permit the longer persistence of rare local endemics by decreasing dispersal rates. However, local endemics may not be among the most common species in a given fragment at the time of its isolation. In this case, they are still more likely to be eliminated than are common metacommunity species that are also more abundant in the fragment when it is isolated from the metacommunity.

In conclusion, the landscape distribution of metacommunity diversity is remarkably complex and rich under the unified theory and ecological drift. Species populations in these model landscapes, particularly under moderate to high dispersal rates, are often fragmented into subpopulations that wax and wane in abundance. In qualitative terms, the patterns of metapopulations in nature look very similar to those in figures 7.5 and 7.6. This is true even though there are absolutely no differences *on a per capita level* among the species. However, *on the species level*, common species are much more competitive because they are more abundant and persistent in the metacommunity, and this gives them a great advantage over rare species under high dispersal rates. Common species simply overwhelm rare species, with the result that rare species are pushed to extinction from the metacommunity under high dispersal rates. The very different fate of common and rare species in the metacommunity under the unified theory underscores once again the fundamental importance of making the neutrality assumption at the individual level, not at the species level (chapter 1).

A final few words are in order about the equilibrium behavior of metacommunity biodiversity under the unified theory. In figures 7.5 and 7.6 it was essentially arbitrary that I considered the local communities as imbedded in a continuous landscape. What was crucially important, however, was the connectivity of local communities or habitat patches by dispersal. Connectivity by dispersal controls the equilibrium species richness and relative species abundance over a fragmented landscape as well. In the simulations presented in figures 7.5 through 7.7, I assumed that dispersal did not connect all local communities equally in the metacommunity. If all the islands or habitat patches in an archipelago are bathed in the same background immigration rate m from the metacommunity as a whole, then the *aggregate* behavior of biodiversity across the archipelago will be indistinguishable from that in a single, unfragmented metacommunity equal to the aggregate size of all the local communities or islands. Indeed, this must be so because the species-area curves for archipelagos look essentially the same as those on the mainland except that they have steeper slopes (MacArthur and Wilson 1967). But these steeper slopes in archipelagos, according to the unified theory, are due primarily to a reduction in the dispersal rate m, rather than to fragmentation per se. Therefore *the critical aspect of fragmentation is less the fragmentation itself, but more its impact on mean dispersal rates and the connectivity of patches or islands, and the interaction of limited dispersal with the stochastic dynamics of biodiversity in the individual patches or local communities.* We explored these questions in greater detail in chapters 5 and 6. Because dispersal limitation is universally present, it is universally true that the precise spatial structure of the habitats in a fragmented landscape and their connectivity will be important (Tilman 1994, Karieva and Wennegren 1995, Harrison 1994, Hurtt and Pacala 1995). In the last few years, there has been a revolution in the making in new mathematical tools for describing patchy landscapes, which often turn out to be fractal (Milne 1991, Ritchie 1997, Richie and Olff 1999). Exploring this burgeoning subject, however, is beyond the limited scope of the present work; but it is clear that there is much room for fruitful development of the theory in these new directions.

SUMMARY

1. The neutral theory says that only a small amount of dispersal connecting the metacommunity is sufficient to maintain presence of a given species in a local community, if the species is reasonably abundant in the metacommunity.
2. The incidence or frequency of a species in a set of local communities increases rapidly with local community size.
3. The covariance in the abundance of a species on two islands of an archipelago is affected not only by immigration from the metacommunity, but also via the exchange of migrants. Covariance is maximal when interisland exchange is large relative to the effect of source-area immigration.
4. Under the neutral theory, the steady-state distribution of biodiversity on the metacommunity landscape is controlled by the fundamental biodiversity number θ and the dispersal rate m.
5. When the dispersal rate is high, populations of individual species are amorphous and diffuse; however, when dispersal rate is low, species tend to occur in more discrete patches of locally high abundance, with high rates of endemism.
6. There is an interaction between dispersal and α and β diversity in local communities and in the metacommunity. When dispersal rates are high, local diversity is high, but metacommunity diversity is low because common metacommunity species wipe out rare endemics. Conversely, when dispersal rate is low, local diversity is lower, with patches of locally common endemics, but metacommunity diversity is much higher. In this case, there is a reduced force of dispersal to displace and cause extinction of the local endemics.
7. The distance decay of similarity in community composition under ecological drift and random dispersal is

expected to be compound exponential. It is compound because abundant and widespread metacommunity species show shallower decay curves than less abundant, more locally distributed species. Distance decay rates are also slower if the metacommunity is linked by high rates of dispersal.

tion event" per birth. Here I will argue that the mere process of asking this question—in the formal context of the unified neutral theory—potentially leads to a profound change in perspective on species, speciation, and the meaning of biodiversity itself. In this chapter I first examine the implications of the neutral theory for phylogeny and phylogenetic reconstruction. I then take a deeper look at the speciation process itself, and the implications of the theory for the nature of species and of biodiversity in general. In this chapter I also examine a second mode of speciation, the "random fission" mode, and show how this alters the equilibrium distribution of relative species abundance in the metacommunity.

Recent advances in molecular biology have opened up new possibilities for understanding phylogenetic relationships far better than ever before. These advances have stimulated renewed theoretical interest in phylogenetic reconstruction (Hey 1991, 1992, Hey et al. 1993, Scotland et al. 1994, Nee et al. 1994, Harvey and Nee 1994). The predictions of the unified theory are qualitatively different in a number of important regards from contemporary phylogenetic theory, however. In current theory, the phylogenetic unit is the *lineage*, and lineages are assigned probabilities of speciating and going extinct—probabilities that have been treated as either constants or stochastic variables by different authors. In the unified neutral theory, however, the phylogenetic unit is the *individual*. Lineages per se do not have assigned probabilities of birth and death. Instead, the stochastic rates of lineage origination and extinction are predicted by the fundamental biodiversity number, θ, and follow inevitably from the distribution of metacommunity relative species abundances. Metacomunity size J_M and the speciation rate ν dictate not only the distribution of lineage lifespans but also the potential of lineages to speciate over their evolutionary lifetimes. Thus, one significant difference from current phylogenetic theory is the fundamental importance attached by the unified theory to the relative abundance of the species that are the lineages of the metacommunity over evolutionary time.

In 1979 marine ecologist Jeff Levinton published an important but insufficiently noticed paper in *Science* that presaged the unified neutral theory of biodiversity and biogeography. Levinton presented a simple verbal model to explain how a diversity equilibrium might arise in the fossil record in the absence of niche assembly. The importance of Levinton (1979) is best understood retrospectively, in my opinion, in the context of the debate that ensued after publication of the "punctuated equilibrium" theory (Eldridge and Gould 1972). Eldridge and Gould argued that the evidence from the fossil record is less consistent with a process of gradual clade diversification in many taxonomic groups, and more consistent with a process of episodic diversification, such that between bouts of rapid cladogenesis, there were relatively long periods of diversity equilibrium. As noted in chapter 1, however, a diversity equilibrium does not necessarily imply a lack of turnover of species (Patzkowsksy and Holland 1997). In any event, this theory stimulated the study of neutral models of randomly branching phylogenetic clades. The idea was to find out whether such models could reproduce the patterns of lineage diversification observed in the fossil record.

Raup et al. (1973) constructed a simple stochastic birth-death process as a neutral model for the branching of phylogenetic trees. They assumed that the per lineage probability of a "birth," or the division of a lineage into two daughter lineages, λ, was the same from one lineage to another, and likewise, that the probability of the death of a lineage, μ, was the same from lineage to lineage. Although total clade extinction is possible when the birth rate exceeds the death rate ($\lambda > \mu$) since it is a stochastic process, the more usual outcome in this case is geometric growth in the number of lineages within clades (Nee et al. 1994). This raises a serious problem with Raup et al.'s (1973) pure birth-death model and other similar models, namely that they do not predict a nonzero diversity steady state (Gould et al. 1977, Stanley 1979, Raup 1985). A diversity equilibrium can be achieved in these models only if additional rules are imposed on how speciation and extinction depend on the number of lineages, a kind of diversity density dependence (Raup et al. 1973, Rosenzweig 1975; see also below).

The general failure of these constant birth-death neutral models to reproduce patterns of steady-state diversity between punctuational events, and other discrepancies between observed and predicted phylogenies, led many to reject neutral models prematurely. Before Levinton (1979), the general conclusion was that the apparent diversity equilibria seen in the fossil record must imply the existence of adaptive community diversity equilibria due to the filling of all available niche space. This means, so the argument goes, that the metacommunity must be niche assembled, and the number of niches sets a limit on total steady-state species richness in the metacommunity. This hypothesis suffers from the same problem that afflicts contemporary, niche-assembly theories of relative species abundance, viz., a free parameter (the number of available niches) that cannot be derived from the theory's first principles.

The significance of Levinton's (1979) paper was to demonstrate that a simple neutral model was fully capable of explaining diversity steady states in the fossil record without invoking any niche assembly rules whatsoever. Imagine a widespread species or taxon that is the stem ancestor to a large clade of descendant species. Assume that this ancestral species and all its contemporary descendant species compete in a zero-sum game for limiting resources. Levinton did not use the term "zero-sum game," but he clearly established an equivalent condition that all occupiable space (or limiting resources) be saturated with organisms. This ancestral species or taxon would have been common with an extensive global range and thus would have been very resistant to extinction due to its very large metapopulation size. Now, start an engine of speciation, by whatever mechanism. New species arise from the ancestral species as well as from species descended from the original stem ancestor. As more and more species are added to the metacommunity, given the zero-sum rule, the average population size of extant species in the clade must decline. As average population size falls, the extinction rate will inevitably rise. A point will be reached when there are so many rare species that the rate of extinction in the clade will increase until it is equal to the rate of origination of new species. At this point the living biodiversity of the clade will increase no further, and a steady-state species richness will be achieved. Levinton's argument was based on average population sizes, not on the steady-state relative abundance distribution of species predicted by the fundamental biodiversity number θ—which of course was unknown at the time—but nonetheless Levinton's conclusion is qualitatively completely correct.

Levinton's model was not the only one to produce an equilibrium diversity. Rosenzweig (1975, 1995) proposed a model of the evolution of continental diversity in which the number of species approaches a steady-state asymptote. Rosenzweig (1995) argued that increased diversity per se will cause increased extinction rates per species. If the unified neutral theory is correct, however, increased extinction rates are simply an incidental by-product of the reduced mean abundances of species when there are more species in the metacommunity. In Rosenzweig's model, unlike Levinton's, the fundamental connection of the diversity equilibrium with limiting resources was not made. Under zero-sum dynamics, average population size must decline with increased species richness. In Levinton's model, as in the unified theory, species per se do not saturate the landscape and limiting resources, *individuals* (collectively) do (unless $\theta = \infty$). Of course, individuals make up species, but from the perspective of the zero-sum game, what matters is the number of individuals, not the number of species.

Rosenzweig also argued that increased diversity will reduce the speciation rate per species. Once again, according to the unified neutral theory, lower speciation rates are not caused by higher diversity but are incidentally correlated negatively with diversity. In the theory, the number of species originating per unit time is controlled by the size of the metacommunity J_M and the per capita speciation rate ν. As the number of species increases for fixed J_M and ν, the speciation rate per species must necessarily decline. Thus, a major difference between the unified neutral theory and previous theories is that *the number of new species arising per unit time is a function of the total number of individuals in the metacommunity, not the number of pre-existing species.* A steady-state diversity is achieved, not because the total metacommunity speciation rate falls. It is achieved because the extinction rate increases until it equals the speciation rate, which is an inevitable outcome of a falling mean population size of metacommunity species under the zero-sum rule.

This perspective puts a different interpretation on the punctuated equilibrium theory itself and, particularly, on the rapid diversification in many taxa that followed mass extinctions in the fossil record. If the unified theory is correct, then a major effect of a mass extinction, beyond killing of a large fraction of the Earth's biota, is to massively desaturate the Earth's resources for a brief period, geologically speaking. Given this brief desaturation, the unified theory predicts that each mass extinction event will be followed by an episode of very rapid population expansion among the relative small number of species that survive the event. During this short nonequilibrium, post-extinction population flush, there will be a very large net excess of births over deaths— many more births per unit time than during periods of diversity equilibrium. In the unified theory, the opportunity for speciation is a function of the expressed birth rate. The theory asserts that it is primarily the temporary but huge surplus of births over deaths that generates the sudden burst of new species following mass extinctions. Such a burst of speciation is predicted *even if there is no change* in the probability of speciation per birth.

Thus, there is no a priori necessity to invoke niche assembly arguments to explain punctuated equilibrium patterns in the fossil record, although adaptive radiation may also occur as a correlated effect of the ecological release accompanying a desaturation of resources. Not all periods of rapid diversification of particular taxa occurred in the aftermath of mass extinctions. Even in these cases, however, the theory suggests that we should seek evidence of regional or global desaturation in the resources limiting the taxon in question precisely during periods of rapid diversification in the taxon. This desaturation of resources can result from the rapid decline or extinction of one or more particularly common metacommunity species in the taxon, or from a sudden increase in limiting resources available to the taxon. The theory also predicts that there should be a "dosage effect," a correlation between the degree of resource desaturation and the amount of subsequent diversification that occurs. Desaturation can also arise from a sudden, massive increase in metacommunity size J_M, which will in turn cause inflation of the fundamental biodiversity number θ. Therefore, diversification will occur with increased J_M even in the complete absence of any change in the probability of speciation per birth ν.

Before discussing the unified theory's predictions about phylogeny, it is useful to outline the current neutral theory of phylogenetic reconstruction. As mentioned, the pioneering work on neutral models in phylogeny was done in the 1970s, particularly by Raup et al. (1973) and Gould et al. (1977). These models have been further developed theoretically by Nee et al. (1992, 1994, 1995), and their predictions have been compared with several molecular phylogenies, particularly of birds (Sibley and Ahlquist 1990), by Harvey and Nee (1994), and Harvey et al. (1994b, 1996). In the case of phylogenies derived from molecular data, the only observable species are extant species, so the full phylogeny as described in the model of Raup et al. (1973), including all its extinct lineages, is unknown. How do one's conclusions differ if one has a "censored" phylogeny, consisting of only its living members? Nee et al. (1994) distinguished four cases based in part on the degree to which the complete or incomplete phylogeny is knowable; figure 8.1 shows a cartoon of the two cases that need concern us. Case 1 represents the full phylogeny from the pure continuous-time birth-death process. In this process, lineages give rise to new lineages at a per lineage rate of λ and go extinct at a per lineage rate of μ, and the clade survives for an arbitrary length of time—the process studied by Raup et al. (1973). Case 2 (case 4 of Nee et al. 1994) is the same phylogeny as case 1, but it retains only those lineages that lead to species alive at the present time T, and from which all extinct lineages

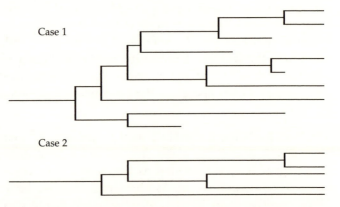

FIG. 8.1. Two cases of phylogenies. Case 1: Phylogeny resulting from the pure birth-death process, with all lineages, living and extinct shown. Case 2: The same clade as in case 1 but with all extinct lineages removed, leaving only lineages alive at the present time T.

are pruned out. Case 2 trees are what Nee et al. (1994) call a *reconstructed phylogeny*, because it is the only phylogeny that can be inferred from molecular data on living taxa. Nee et al. (1994) make the assumption that all reconstructed phylogenies are monophyletic and that reticulate evolution (e.g., allopolyploidy) is unimportant (but see Arnold 1997).

Following Nee et al. (1994), let $\Pr\{i, t\}$ be the probability that a clade has i lineages at time t, and let $\Pr\{t, T\}$ be the probability that a single lineage alive at time t has at least one surviving lineage at present (time T). Now consider the pure birth-death process (case 1). $\Pr\{t, T\}$ is equal to:

$$\Pr\{t, T\} = \frac{\lambda - \mu}{\lambda - \mu \exp\{-(\lambda - \mu)(T - t)\}}.$$

Thus, $\Pr\{0, T\}$ is the probability that a clade starting with a single lineage at time 0 is not extinct at time T. The probability that a clade starting with a single lineage is extinct at time t is therefore

$$\Pr\{0, t\} = 1 - P(0, t),$$

and the probability that the clade has i lineages at time t is given by

$$\Pr\{i, t\} = \Pr\{0, t\}(1 - u_t)u_t^{i-1}, \qquad i > 0,$$

where

$$u_t = \frac{\lambda\{1 - \exp[-(\lambda - \mu)t]\}}{\lambda - \mu \exp[-(\lambda - \mu)t]}$$

is the probability of a new lineage arising in the time interval 0 to t. $\Pr\{i, t\}$ has a simple geometric distribution (Kendall 1948).

The pure birth-death process cannot be observed directly because of past extinctions within the clade, so we need to derive the equivalent expressions for the "censored" phylogeny consisting of the lineages surviving to the present. Nee et al. (1994) show that the probability of i observed lineages surviving to time T (case 2) is also geometrically distributed with a different geometric parameter: u_t' of the pure birth-death process (case 1) weighted by the ratio of the probabilities of a single lineage surviving to the present (time T) and to intermediate time t:

$$\Pr\{i, t, T\} = (1 - u_t')(u_t')^{i-1}, \qquad i > 0,$$

$$\text{where } u_t' = u_t \frac{P(0, T)}{P(0, t)}.$$

Note that there is no zero term in the expression for $\Pr\{i, t, T\}$. This is because at least one lineage will survive to the present. Otherwise we generally would not know of the clade's existence; but even if it were known, we would not have the opportunity to collect molecular data from it.

The predicted shape of the frequency distribution of number of the descendant lineages from the geometric model of Nee et al. (1994) is shown in figure 8.2. In all cases, the most frequent category represents lineages with only one living descendant (themselves), and the frequency of lineages with a higher number of descendants falls off exponentially. The longer the time period sampled, the larger the number of possible descendant lineages, and the frequency distribution becomes flatter; but the highest frequency category remains the singleton category.

Some phylogenies have distributions like figure 8.2, but many do not. Some lineages are far too "bushy" for the geometric distribution. Harvey and Nee (1995) and Harvey et al. (1995, 1996) conclude that these nonfitting phylogenies are inconsistent with neutral cladogenesis. According to the unified theory, however, this conclusion is premature. The geometric distribution is *not* the distribution predicted by the unified neutral theory—except in cases of lineage-poor clades or small time samples. If the time window is too short or if the taxonomic range is too narrow, then most of the clade will not be sampled, and the distribution will

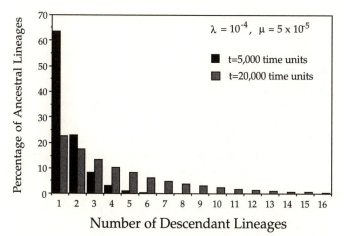

FIG. 8.2. The geometric distribution of the number of descendant lineages expected under pure birth-death process of Nee et al. (1994). Arbitrary time units birth rate $\lambda = 0.0001$ and death rate $\mu = 0.00005$. *Black-bar histogram:* Phylogeny sampled over a period of 5000 time units. *Gray-bar histogram:* Phylogeny sampled over 20,000 time units. Note flattening of the distribution as the sampling interval increases, as the frequency of larger numbers of descendant lineages increases. In the geometric model, the category of one descendant lineage is always the most frequent.

appear to be geometric-like. However, the unified theory predicts that as longer and longer time windows are considered, an interior mode will appear in the distribution of number of descendant lineages, and the most abundant category will no longer be singletons.

This effect is superficially analogous to the "veil line" sampling phenomenon discovered by Preston (1948) in the distribution of relative species abundance (chapter 2). In spite of superficial similarity, however, the interior mode in the distribution of the number of descendant lineages has a different explanation than the interior mode of the relative species abundance distribution (chapter 5). The unified theory's explanation follows from the prediction that individual lineages must have very different probabilities of

241

speciating and going extinct because they differ enormously in their abundances over evolutionary time. Species that are abundant and widespread throughout the metacommunity have very long expected lifespans measured in geological time, far longer than the lifespans of rare and local species. Besides having much shorter lifespans, rare lineages have fewer opportunities to speciate also because they produce absolutely fewer offspring per unit time, even though their per capita birth rate is the same. Therefore, common species will be the ancestors of more species and of more present-day biodiversity than rare species.

The actual distribution of the number of descendant lineages will also be affected by the frequency distribution of abundances of ancestral lineages. From metacommunity relative species abundance under point mutation speciation, we know that there are fewer common lineages than lineages of intermediate abundance, and fewer intermediate-abundance lineages than rare lineages. Multiplying longevity (which is a function of lineage abundance) by the number of lineages of a given abundance results in most speciation events occurring in ancestral lineages having intermediate abundances. This is the explanation for the interior mode of the distribution of the number of descendant lineages. An interior mode is also found under random fission speciation, but the metacommunity relative abundance distribution in this case is zero-sum multinomial, so that rare species are less frequent and even less likely to be the ancestors of many descendant lineages (see below).

Distributions of the expected number of descendant lineages are shown in figure 8.3 as a function of the abundances of the ancestral lineages and of the fundamental biodiversity number, θ. The distributions are negatively skewed, with most of the daughter lineages (percentage of speciation events) coming from abundant ancestral lineages. When θ is small, there is high dominance and low diversity in the metacommunity, and the distribution of speciation events

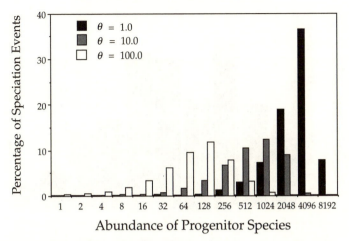

FIG. 8.3. Distribution of the number of daughter lineages at steady-state diversity equilibrium in the metacommunity, as a function of the fundamental biodiversity number θ. Note the negative skew of the distributions and the concentration of daughter lineages in high-abundance ancestors.

is right shifted. As θ increases, there are more metacommunity species of intermediate abundance, so the distributions become broader and lower. However, the distributions remain negatively skewed, such that most of the steady-state diversity always originates from the more abundant ancestral species, not from the rarer species.

As an empirical example, the phylogeny of birds represented by the distribution of species per monophyletic family clearly has an interior mode (fig. 8.4). These data were tabulated by Sibley and Ahlquist (1990) and include 8501 of the estimated 9700 extant species of birds, and more species than the 1700 species included in Sibley and Ahlquist's partial molecular phylogeny of the birds.

Herein, in my opinion, lies a fundamental conceptual weakness in the neutral models of Raup et al. (1973) and Nee et al. (1994, 1995)—and all other current models of phylogeny of which I am aware—namely, their failure to

243

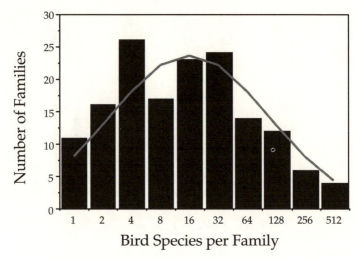

FIG. 8.4. Distribution of the number of species per family of birds, for 8501 of the approximately 9700 species of birds in the world's avifauna. Data compiled by Sibley and Ahlquist (1990). The line is the best-fit lognormal.

take the relative abundance of lineages into account. Nee et al.'s results were derived under the assumption that λ and μ are probabilities that can be measured and defined on a per lineage basis. In contrast, the argument of Levinton (1979) and the unified theory is that the instantaneous per lineage probabilities of speciation and extinction are dictated by the current relative abundances of lineages. The omission of relative abundance is surprising because, at least in demography theory, population size has long been recognized as *the* critical parameter controlling the expected time to extinction (Levins 1970, Richter-Dyn and Goel 1972, Lande 1988, 1993).

One possible explanation for this omission is that paleobiology utilizes species presence-absence data more often than data on relative species abundance. Thus, the importance of relative abundance is perhaps less widely appreciated in paleobiology than in population and community ecology. It may also be that the quality of relative abundance data in the fossil record is more difficult to assess. However, I suspect that the data have simply not been collected in many cases. Indeed, data on relative lineage abundance through time in the fossil record might actually be a better predictor of relative lineage longevity than the short-term data on modern relative species abundance available to ecologists.

In the unified neutral theory, we can study analytically how factoring in relative lineage abundance affects the theory of phylogeny. First we need to compute the expected lifespans of species undergoing zero-sum drift in the metacommunity as a function of their initial abundance. This problem is closely related to the "absorbing case" problem that we studied in chapter 4, namely the expected time that it takes a species undergoing zero-sum ecological drift to go locally extinct or achieve monodominance. In the present metacommunity problem, the absorbing state of monodominance no longer exists because of the low but finite probability of a speciation event. The only remaining absorbing state is global extinction of the species from the entire metacommunity. Therefore the matrix of transition probabilities is identical to the absorbing case studied in chapter 4, with the exception that

$$\Pr\{J - 1 | J\} = \nu$$

and

$$\Pr\{J | J\} = 1 - \nu.$$

That is, the probability per birth that a monodominant species in the metacommunity will be dethroned from monodominance is the probability of a speciation event ν.

With this change, we can now calculate exactly the expected lifespan of arbitrary species i at initial abundance N_i, where lifespan is measured in total number of deaths in the community until extinction of the ith species, using the same method detailed in chapter 4. The mean lifespan Ω of species i starting with N_i individuals is

$$\Omega(N_i) = \sum_{k=1}^{N_i-1} k \frac{\binom{J_M}{k}}{\binom{J_M-2}{k-1}} + N_i \left[\sum_{k=N_i}^{J_M-1} \frac{\binom{J_M}{k}}{\binom{J_M-2}{k-1}} + \frac{1}{\nu} \right].$$

The first summation is the number of times that the ith species, starting at abundance N_i, will pass through abundances 1 through $N_i - 1$ before extinction, and the second summation is the number of times that the ith species will pass through abundances N_i through $J_M - 1$. The final term, N_i/ν, is the time ith species will spend being monodominant before extinction.

Note that the dominant term in $\Omega(N_i)$ is N_i/ν, the time the ith species spends in the monodominant state. This is because leaving the state of monodominance must await a speciation event, which can be a very long wait indeed. Because this is so very much longer than the residency times at other abundances, and because real species have a vanishingly small probability of ever achieving monodominance in the entire metacommunity, at least for large metacommunities, I will henceforth ignore this term. Figure 8.5 shows how the initial abundance of a species affects its residency time at each abundance $1 \leq N_i \leq J_M - 1$ before the species goes extinct. I have illustrated the behavior of $\Omega(N_i)$ for a very small metacommunity ($J_M = 16$), but the results are qualitatively the same for all larger metacommunity sizes. Species starting their existence at small population sizes have short residency times at all abundances. Species starting at high abundance have progressively longer residency times at higher abundances.

One of the most interesting special cases is the extinction time of newly arisen species. Under the point mutation mode of speciation, calculated as probability ν per birth, the

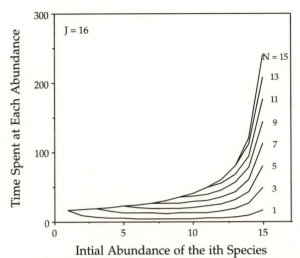

FIG. 8.5. Time spent at each abundance (residency time), short of complete monodominance, as a function of the initial abundance of the ith species, in a community of size 16.

expected time to extinction (total number of deaths in the metacommunity) of a newly originated species is (ignoring monodominance)

$$\Omega(1) \cong \sum_{k=1}^{J_M-1} \frac{\binom{J_M}{k}}{\binom{J_M-2}{k-1}}.$$

As noted in chapter 4 regarding time to fixation, calling $\Omega(N_i)$ an "extinction time" is a bit misleading because it actually represents the number of deaths in the metacommunity before a species starting with abundance N_i goes extinct. $\Omega(N_i)$ counts deaths in *all* species, not just deaths in species i. The value of $\Omega(N_i)$ increases with metacommunity size mainly because more deaths of all species occur per unit time in a large metacommunity than in a small one.

247

Converting the number of deaths until extinction to an absolute time scale is actually straightforward. All we need to do is normalize for metacommunity size. For example, on average twice as many deaths occur per unit time in a metacommunity that is twice as large. Therefore, to put extinction times in these two communities on the same absolute timescale, simply divide the number of deaths by two in the community that is twice as large. After normalization, however, one will still find that a species starting from a given abundance will live longer in a larger metacommunity. This is because a species starting with N_i individuals in a larger metacommunity has potentially more states of abundance that it can pass through before extinction occurs. Figure 8.6 shows how total time to extinction of the ith species depends on initial population size and the size of the metacommunity. In this figure I have scaled all extinction times on the same timescale, relative to the number of deaths in the smallest community ($J_M = 16$). Thus, the total number of

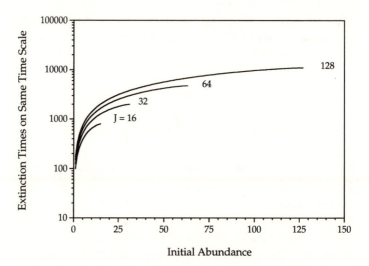

Fig. 8.6. Scaled extinction times for metacommunities of different sizes, as a function of the intial abundance of the ith species.

248

deaths before extinction of the ith species for the community of size 32 is divided by 2; the total for the community of size 64 is divided by 4, and so on.

We are now in a position to calculate the distribution of lifespans expected at equilibrium between speciation and extinction in the metacommunity. Recall from chapter 5 that the expected metacommunity relative abundance distribution is given by

$$E\{r_i|J_M\} = \sum_{k=1}^{C} r_i(k) \cdot \Pr\{S, r_1, r_2, \ldots, r_S, 0, 0, \ldots, 0\}_k,$$

where C is the total number of configurations, $r_i(k)$ is the abundance of the ith ranked species in the kth configuration, and $\Pr\{S, r_1, r_2, \ldots, r_S, 0, 0, \ldots, 0\}_k$ is the probability of the kth configuration. The metacommunity dominance-diversity curve is the set of ordered expectations, $E\{r_i\}, i = 1, 2, \ldots$, ordered such that species of the lowest rank are the commonest. Therefore, the equilibrium distribution of lifespans is the distribution of

$$\Omega\big(E\{r_i\}\big), i = 1, 2, \ldots.$$

In general, the $E\{r_i\}$ are nonintegral expectations. To calculate this distribution, one can use the gamma function to calculate nonintegral factorials, at least for species whose expected abundances are greater than unity. I prefer an alternative, however. In chapter 9, I present the recipe for stochastically simulating the metacommunity relative abundance distribution. Each simulation results in only integer abundances of each species. For each simulation one can compute $\Omega(N_r)$ exactly, where N_r is the observed abundance of the rth ranked species in the given simulation. Then, to obtain arbitrarily accurate estimates of the distribution of lifespans, one averages across the ensemble of simulation results. Accuracy increases as a function of sample size, the number of simulation runs.

249

Interestingly, the fundamental biodiversity number θ does not uniquely determine the distribution of species lifespans, i.e., the distribution of Ω, even though θ does uniquely determine the steady-state metacommunity distribution of relative species abundance. However, the two parameters that compose θ, metacommunity size J_M and speciation rate ν, both separately enter the function for Ω.

I now compute the equilibrium distribution of lifespans of all metacommunity species in a metacommunity of size $J_M = 100,000$. Figure 8.7 displays what might be termed *lifespan-diversity curves* in analogy to dominance-diversity curves. The x-axis is species rank in abundance, as in a standard dominance-diversity curve. However, the y-axis is the common logarithm of the mean lifespan for species of a

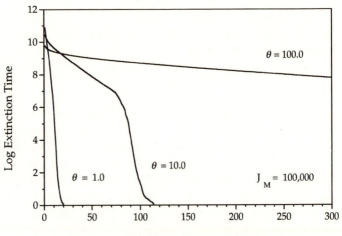

Fig. 8.7. Lifespan-diversity curves measured in total deaths in a meta-community J_M of size 100,000, for various values of the fundamental biodiversity number θ. Means of 1000 simulations. The breaks in the curves occur at the rank above which all communities cease having the same number of species. As $J_M \to \infty$, the curves approach straight lines as in the logseries for infinite metacommunity size.

given rank in abundance in the metacommunity. The curves are for three values of the fundamental biodiversity number θ: 1, 10, and 100. As $J_M \to \infty$, the curves approach a logseries-like straight line on the semilog plot of log extinction time versus species rank, and the downward inflection point disappears. This downward bend in the curve is found in finite metacommunity sizes and arises because not all communities of a given size have the same number of species. The larger the metacommunity, the higher the rank at which the inflection point occurs, for a given θ value. Note that increasing metacommunity diversity actually decreases the lifespans of the most common species, although it increases the lifespans of species of intermediate abundance (fig. 8.7).

The most important conceptual point made by Figure 8.7 is that there are enormous differences in longevity among the metacommunity species, ranging over eleven orders of magnitude even in this relatively small metacommunity of size $J_M = 100,000$. For metacommunities of sizes in the billions or trillions, the curves are much higher on the y-axis, and the longevities range over far greater ranges even than those in figure 8.7. These huge differences in persistence time of common and rare metacommunity species have significant consequences for phylogeny, particularly for the number of daughter lineages that each metacommunity species produces.

We can use the theory for the absorbing process of ecological drift for the ith species developed in chapter 4 to count the number of speciation events that are expected in a given lineage before the lineage goes extinct. To do this, we need to add additional states and transition probabilities to the Markovian model for the absorbing process. Let a new species arise as a single individual with probability ν

per birth, and let N_j be the new species. Thus,

$$\Pr\{N_i - 1, N_j = 1 | N_i\} = \left(\tfrac{N_i}{J}\right)\nu$$

$$\Pr\{N_i - 1, N_j = 0 | N_i\} = \left(\tfrac{N_i}{J}\right)(1 - \nu)\left(\tfrac{J - N_i}{J - 1}\right)$$

$$\Pr\{N_i | N_i\} = \left(\tfrac{N_i}{J}\right)(1 - \nu)\left(\tfrac{N_i - 1}{J - 1}\right)$$

$$+ \left(\tfrac{J - N_i}{J}\right)\left(\tfrac{J - N_i + 1}{J - 1}\right)$$

$$\Pr\{N_i + 1 | N_i\} = \left(\tfrac{J - N_i}{J}\right)\left(\tfrac{N_i}{J - 1}\right)$$

In this system, monodominance is no longer an absorbing state because the monodominant species can be invaded by the new species. In the $D = 1$ case, note that a new species can be generated only when the ith species declines in abundance by one individual (the point mutation version of speciation). The first equation is the transition probability for a speciation event. The second equation is the transition probability for a decline in abundance but no speciation event in the ith species.

In these equations, we are counting only new species whose ancestor is species i. We must do this counting while simultaneously allowing the ith species to take its zero-sum random walk, so we need to keep track of the abundance of the ith species as speciation events occur. This means that we need $2J + 1$ states that record both the current abundance of the ith species as well as whether a speciation event has or has not just occurred. There are only J additional states, not $J + 1$, because the absorbing state of zero abundance cannot produce a new species.

We can now solve for the number of times that the ith species gives rise to a new species as a function of the abundance of species i. Once again ignoring the monodominant state, the number of daughter species Λ produced over the evolutionary lifetime of species i having initial abundance N_i is

$$\Lambda(N_i) = \frac{\nu}{J_M} \sum_{k=1}^{N_i-1} k^2 \frac{\binom{J_m}{k}}{\binom{J_M-2}{k-1}} + \frac{N_i\nu}{J_M} \sum_{k=N_i}^{J_M-1} k \frac{\binom{J_M}{k}}{\binom{J_M-2}{k-1}}.$$

From $\Lambda(N_i)$, it is clear that the likelihood that any given metacommunity species will produce a daughter species is small to very small. This is because the speciation rate and metacommunity size enter $\Lambda(N_i)$ as the *ratio* of a very small number to a very large number, ν/J_M. Nevertheless, because of the enormous differences in longevities among species, common species will produce most of the new species that do originate. We have already demonstrated this result and seen the behavior of $\Lambda(N_i)$ in figure 8.3.

These results have immediate application to the reconstruction of phylogenies. The appearance of clades under the unified theory is quite different from those that arise under the models of Raup et al. (1973) and Nee et al. (1994). Consider, for example, a cladogram produced by the Raup et al. model (fig. 8.8). In this model, recall that if the probability of birth λ is greater than the probability of death μ of a lineage, then the usual outcome is exponential growth of the number of lineages. This is true for the example illustrated, a case in which λ was set equal to $9\cdot10^{-5}$ and μ was set to $4\cdot10^{-5}$. In figure 8.8, I have illustrated Nee et al.'s case 4(case 2 in fig. 8.1), showing only the "censored" phylogeny of lineages surviving to the present (right edge). Note that case 2 cladograms always make it appear that the speciation rate is increasing with time. This is a visual artifact of pruning out all extinct lineages. In actuality, the probability of lineage birth and death has remained constant through time. Nevertheless, there is a real geometric increase in the number of lineages.

I now illustrate two cladograms produced by the unified theory, under the point mutation mode of speciation. The first is an example of a cladogram produced for a value of

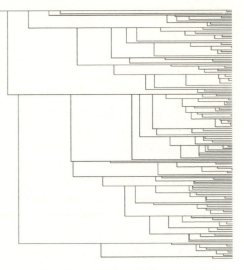

FIG. 8.8. A typical case 2 cladogram produced by the stochastic phylogeny models of Raup et al. (1973) and Nee et al. (1994), which assume constant per lineage probabilities of birth λ and death μ. In this example, $\lambda = 9 \times 10^{-5}$ and $\mu = 4 \times 10^{-5}$. Note the geometric increase in number of lineages. This is partly an artifact of pruning out all extinct lineages.

the fundamental biodiversity number θ of 100, again showing only extant lineages (fig. 8.9). Several differences from figure 8.8 are immediately apparent. First, lineages and biodiversity are much more concentrated at the present time and in the relatively recent past. Once again, this is a case 2 reconstructed phylogeny, so part of this effect is due to pruning out the extinct lineages. But mostly this effect is due to the fact that new lineages arise as rare species that have a very high probability of rapid extinction. Second, there are many more old lineages extending much farther back in time than in the Raup model (figure 8.8). This is not because the clade is polyphyletic, but because a number of very abundant lineages are very long lived. All share a common ancestor and diverged from it long before the time

254

FIG. 8.9. A typical case 2 cladogram produced by the unified theory for a fundamental biodiversity number $\theta = 100$. This is for the point mutation mode of speciation. Note the concentration of most diversity at the present and recent past. This is partly due to the pruning out of extinct lineages. However, in the unified theory under point mutation speciation, the diversity really is more concentrated near the present because of rapid extinction of new but rare species.

horizon shown in figure 8.9. Third, some clades are far more bushy and speciose than others. This is because very abundant metacommunity species are ancestors of more daughter lineages than less abundant lineages. This happens for two reasons. First, speciation occurs on a per birth basis, not a per lineage basis, and many more births occur per unit time in common species than in rare ones. Therefore, common species are more often the progenitors of new species. Second, by far the most important quantitative factor is that very common metacommunity species are extremely long lived and old (quite literally "older than the hills"). For this reason they are sources for new species in the long run far more than short-lived rare species. The final difference to note is that a steady-state biodiversity is achieved in the phy-logenetic clade. Even though it looks like diversity is increas-ing, the present-time biodiversity is actually in equilibrium because the rate of speciation is in stochastic balance with the rate of extinction at all times. Reading backwards in time from the present (right to left in fig. 8.9), after most of the newly arisen species go extinct, there is a much slower decay of biodiversity in the clade. This is because these ancient lineages are the very abundant metacommunity species that are very resistant to extinction.

The biodiversity equilibrium achieved in phylogenetic clades is strongly dependent on the biodiversity number θ as well as on the mode of speciation. I defer considera-tion of the phylogenetic pattern that arises under the ran-dom fission mode until I discuss speciation later in this chapter. However, I now illustrate what happens when we use a smaller value of the fundamental biodiversity num-ber, $\theta = 10$. From the smaller θ, we should expect a reduc-tion in equilibrium clade diversity, and this is in fact what happens (fig. 8.10). There are at least two other interesting consequences. First, because of zero-sum dynamics, we know that the common species in a lower diversity metacommu-nity will be more abundant than in a species-rich metacom-munity. Therefore, *the proportion of long-lived lineages out of all living lineages in the clade should be higher in metacommuni-ties with a smaller θ*. This is because new species arise more slowly, and previously extant species are therefore, on aver-age, more abundant. Second, a corollary conclusion is that *the average age of most lineages will be older in the metacommunity with a smaller θ*. Neither of these predictions emerges from the Raup et al. (1973) models.

I conclude this section by discussing a sampling issue that is similar to the "veil line" problem posed by Preston (1948, 1962) in the sampling of relative species abundance (chapter 2). What happens to the apparent phylogeny if we do not find all the species? This is an important problem

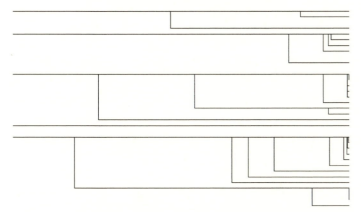

FIG. 8.10. A typical case 2 cladogram produced by the unified theory, for a fundamental biodiversity number $\theta = 10$. This is for the point mutation mode of speciation. Note that the average age of species increases with a reduction in θ, and the proportion of long-lived lineages increases. These effects occur because the species in a species-poor metacommunity are more abundant than in a species-rich community because of zero-sum dynamics.

not only for estimating modern living biodiversity on Earth, but also for the classical problem in paleontology of estimating the number of species that have ever lived and therefore the true speciation rate. The essential point to make, which was Preston's point as well, is that we are not equally likely to discover each and every extant lineage or species. Common and widespread species will be found, but as species become rarer and rarer, they are more and more likely to be overlooked. Thus, the sampling of species—both in the modern world and in the fossil record—is heavily biased toward finding the spatially and temporally extremely abundant species, a well-known problem to paleontologists and ecologists alike. This bias is a huge problem for the fossil record, but it is also a severe problem for contemporary biodiversity studies.

I illustrate this sampling problem by showing the cladogram that one obtains if the rare species are not known.

FIG. 8.11.   Same phylogeny as in figure 8.9, with $\theta = 100$, except that all rare lineages have been pruned out. This increases the apparent age of species and reduces the apparent diversity of clades at the present.

In figure 8.11, I present the same phylogeny that was shown in figure 8.9, except that now all of the rare species (<32 individuals in this numerical example) have been eliminated. Two things are immediately apparent from figure 8.11. First, the terminal bushiness of the clades at the present time is much reduced. This is because most of the rare species are recent in origin. Fewer old species are rare because rare species do not tend to persist. Second, the average age of lineages appears to be older. This age bias is because the species left in the clade after removing the rare species are more resistant to extinction and therefore more persistent. An important conclusion from the unified theory, therefore, is that *a sampling bias toward common species causes an upward bias in the estimate of the mean age of species.* As rarer and rarer species are collected from a given clade, the estimated mean age of species in the clade will inevitably decrease. The neutral theory affords us a direct method to quantify the magnitude of this bias. We know the theoretical longevities of metacommunity species from the distribution of $\Omega(N_i)$, so we can simply compute the expected mean lifespans for any threshold abundance N_i and above.

The importance of understanding this bias is more than simply estimating the ages of lineages; it also leads us to a systematic underestimation of the speciation rate. I will have more to say on this topic below.

Thus far I have not directly addressed the speciation process and the deeper meaning of biodiversity itself, namely: What is it that we are really enumerating when we count species in the metacommunity or construct a phylogeny? The unified neutral theory gives a surprising and potentially challenging answer to this question that generates a whole array of additional questions for testing in the future. In chapters 2 and 3, I discussed the almost infinite regress of rarity found in observed relative species abundance distributions. As sample sizes increase, the rarest species become ever rarer relative to the commonest species, so that the range of abundances grows larger and larger. We now know that this is a necessary result from the equilibrium theory of metacommunity biodiversity presented in chapter 5. The unavoidable conclusion is that *the more extensively and finely we look, the more biodiversity we find.*

The second conclusion we can draw from observations of biodiversity is that if we lump species into related clades, such as species into genera and genera into families, *the qualitative pattern of abundance and diversity among the higher taxa remains fundamentally unchanged.* We have already illustrated this phenomenon for families of birds (fig. 8.4). A second example showing this pattern is the distribution of plant species per family on three different spatial scales: BCI, Panama, and the entire world (fig. 8.12). These distributions are all very much like the zero-sum multinomials that characterize the distribution of individuals per species (chapters 3, 5),

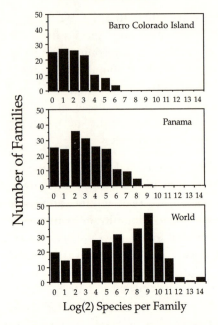

FIG. 8.12. The distribution of species per family of flowering plants, as a Preston-type plot on three different spatial scales—BCI, all of Panama, and the world. Shown is the veil line sampling phenomenon first identified by Preston, and also the long tail of species-poor families, very reminiscent of the zero-sum multinomial distribution of relative species abundances.

for example, the distribution of the abundances of breeding birds of Great Britain (fig. 2.6).

This self-similarity of the organization of biodiversity on different scales of taxonomic resolution leads potentially to a radical departure from the typological view of species because it implies that *biodiversity is intrinsically and fundamentally fractal.* The unified neutral theory provides strong theoretical support for the hypothesis of the fractal nature of biodiversity. Phylogenetic clades generated by the theory have fractal geometry. To see this, perform a sampling of a given clade by counting the number of surviving

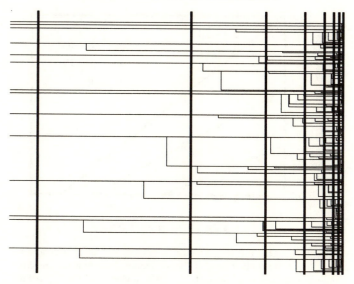

FIG. 8.13. Sampling of a phylogeny produced by the unified neutral theory for measuring its fractal dimension. The number of lineages is counted at exponentially deeper sampling times into the past, and then the logarithm of the number of lineages is plotted against the logarithm of the time depth. This phylogeny was produced by a fundamental biodiversity number θ of 100.

lineages at exponentially increasing sample times into the past (fig. 8.13). Then plot the log of the number of lineages as a function of the log of the time depth of the sampling points on the phylogeny. If these cladograms are self-similar and homogeneous fractals, then these log-log plots should be linear with a slope $-D$, which is the fractal dimension of the phylogeny. The phylogenies predicted by the theory are not only fractal, but the fractal dimension of the phylogeny bears a functional relationship to the fundamental biodiversity number, θ, under the point mutation mode of speciation (fig. 8.14). Under the random fission mode of speciation, the fractal dimension is jointly related both to the speciation rate ν and the size of the metacommunity J_M.

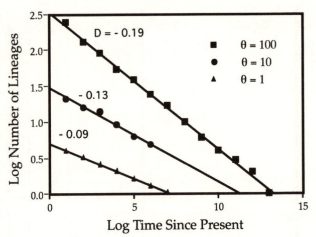

FIG. 8.14. The fractal geometry of phylogenies and of biodiversity under the unified neutral theory. Figure shows the relationship between the log of number of lineages in a clade and the log of the time since the present that the clade is sampled. The fractal dimension of the phylogeny is functionally related to the fundamental biodiversity number θ. Each line is fit to the data from one phylogeny, for $\theta = 100$, 10 or 1. They are not ensemble means of many runs. I chose to illustrate single runs to show the precision of the relationship for single phylogenies. The fractal dimension of a phylogeny decreases with decreasing θ.

This is because these parameters become separated in their effect on metacommunity diversity in the theory for random fission speciation (see below).

If the neutral theory's assertion is correct that biodiversity is fundamentally a fractal, then this conclusion has implications for the way that we think about and count species. If the self-similarity of biodiversity extends smoothly all the way down to the individual level, then the fractal geometry of biodiversity would be homogeneous at all taxonomic scales. But if this is so, where is the biodiversity signature of the existence of species? A signature of species would consist of evidence that the fractal geometry of biodiversity changes and that a different scaling region exists below the species level. A change in the scaling region would imply that the processes generating diversity are fundamentally different above and below the species level. Such a change in scaling region would be manifest as a "kink" or bend in the lines describing the fractal geometry of biodiversity (fig. 8.14) at the species level. If no change in slope occurs at the species level, however, then the geometry of biodiversity will provide no evidence for (or against) the special nature of species, even if species have biological reality in other terms, such as in terms of genetic isolation.

The unified neutral theory is silent on the existence of species and their nature. The theory predicts that biodiversity should be a homogeneous fractal with one scaling domain all the way down to the individual level. If biodiversity is a perfectly homogeneous fractal, and if there are no kinks in the fractal geometry at the species level, this implies that a satisfactory answer to the question—how many species are there?—cannot really be answered except operationally. It requires a definition of the scale of aggregation of biodiversity that we call species. This scale is unlikely to be totally arbitrary, even if diversification below the species level is continuous down to the individual level, as the theory asserts. However, this species scale will then be determined less by the true fractal nature of biodiversity, than by the average difference that separates whatever taxonomists identify as "good" species. But whatever we decide is a "good" species, an empirical question remains, namely: To what extent is biodiversity a homogeneous fractal all the way up and down the ladder of taxonomic scale? Even if biodiversity is truly fractal, as the neutral theory asserts, it does not have to have the same self-similarity across all scales. This question and its ramifications are likely to be very fruitful areas for research by ecologists and evolutionary geneticists for years to come.

Up to this point in the book I have explored the neutral theory mainly under the assumption of one particular mode of speciation, a mode in which new species arise like rare point mutations. I now consider some consequences of changing the mode of speciation to one in which species arise by the random fission of an ancestral species into two daughter species. The decision to explore other modes of speciation was made relatively late in the writing of this book (with the help of my graduate seminar in biogeography), so I have not yet developed the theory for other modes of speciation as fully as would be desirable. A more complete neutral theory for alternative modes of speciation must therefore be set aside for later work. Nevertheless, my results so far suggest that pursuing the implications of different modes of speciation for biodiversity and biogeography would be a valuable and important exercise.

The random fission mode of speciation is an attempt to capture the essence of allopatric speciation (Mayr 1963). The Mayrian concept of species is one of populations reproductively isolated by a complex of pre- and postmating isolating mechanisms. These barriers to the free exchange of genes are hypothesized to arise first in allopatry as correlated responses to selection in the different environments experienced in allopatry. These barriers are believed to be reinforced by selection against hybrids once secondary contact between the populations is established. Partial or total reproductive isolation between species impedes the free flow of genetic information, and this should leave a signature in the fractal geometry of biodiversity.

Whether we fully accept this model of speciation, we can still ask: What will the biodiversity signature of Mayrian species be? Specifically, what impact would the allopatric origination model have upon the metacommunity equilibrium relative species abundance distribution? If we assume that allopatry is required for speciation, then one of the major differences between this mode of speciation and the point mutation mode studied earlier is that incipient allopatric species would usually start their existence at some high or moderately high initial abundance. The consequences of being common at origination are potentially profound for metacommunity biodiversity because new species will not go extinct as rapidly as they do under the point mutation mode of speciation. Alternatively, it may be that most allopatric speciation events occur in small peripheral isolates, in which case the random fission model may actually be less appropriate than the point mutation model, even though the speciation event occurred in allopatry. From the theoretical point of view, the critical issue is not allopatry per se, but the size of the species population at its origination.

From the perspective of the unified theory, the physical nature of the barrier causing the allopatry is unimportant. The essential point is that an ancestral species population is split into two daughter species. I assume that splitting takes place randomly and cuts the ancestral population into two generally unequal fragments, like the single cut of a knife. I assume that this is a random uniform process, equally likely to cut the population into any two sizes summing to the ancestral population size. This is analogous to a geographic barrier randomly dividing the ancestral population. From the point mutation mode I retain the definition of speciation rates on a per birth basis. Thus, an individual is picked at random from the metacommunity and its species is determined. I then let this species undergo random fission into two daughter species. In order to make the bookkeeping analogous to the point mutation mode, I also assume that the ancestral species persists as the larger population of the two daughter species. This avoids an accounting problem of pseudo-extinction—false extinctions counted when a species evolves into its own descendant.

The random fission mode of speciation has a dramatic effect on metacommunity diversity. In figure 8.15, I compare the equilibria for point mutation versus random fission for a metacommunity of size = 10,000 and a fundamental biodiversity number $\theta = 10$. The equilibrium number of species

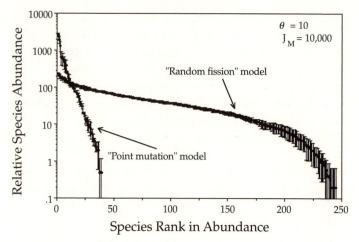

FIG. 8.15. Comparison of the equilibrium distribution of relative species abundance in the metacommunity for the point mutation mode of speciation versus the random fission mode, for a metacommunity size $J_M = 10,000$, and $\theta = 10$. Means ± 1 standard deviation.

is increased by nearly a factor of six in this numerical example. The reason for this large increase in species richness under random fission is the longer average lifespans of new species due to their larger population sizes at origination.

Although it may not be apparent from the dominance diversity curve shown in figure 8.15, there is also a qualitative change in the functional shape of the distribution. It turns out that the metacommunity distribution under random fission speciation is not the logseries, but rather it is a zero-sum multinomial with an interior mode (fig. 8.16). The interior mode arises because rare species are not at as high a frequency as under the point mutation mode. In that case, the high frequency of rare species was maintained by the continual input of new rare species, all of which originate at extreme rarity as singletons.

Another important difference between the two modes of speciation is that the fundamental biodiversity number θ

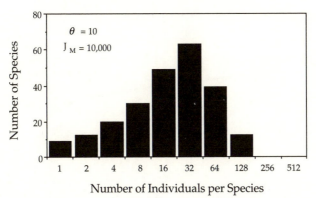

FIG. 8.16. An example of a zero-sum multinomial that arises as the equilibrium distribution of relative species abundance in the metacommunity under the random fission mode of speciation. The interior mode differentiates this distribution from the one predicted by the point mutation mode, which yields a logseries in the infinite J_M limit.

no longer uniquely determines the metacommunity distribution of relative species abundance under the random fission mode. The difference is that both metacommunity size J_M and the speciation rate ν have to be specified separately because their effects are no longer the same. Although the distribution is uniquely determined by J_M and ν, these two parameters are decoupled in the functions yielding the probability distributions of relative species abundance in the metacommunity. This decoupling can be proven analytically by solving for the equilibrium distribution of abundances for small J_M. I have not yet been able to find the analytical solution for arbitrarily large J_M, however.

The results presented in figure 8.17 show that θ does not uniquely determine the metacommunity distribution under the random fission mode. In all four cases, $\theta = 4$, but the metacommunity sizes and speciation rates are both being varied. The reasons why J_M and ν are decoupled in the random fission mode of speciation are easy to discern. When new species do not always arise as singletons, but can be any

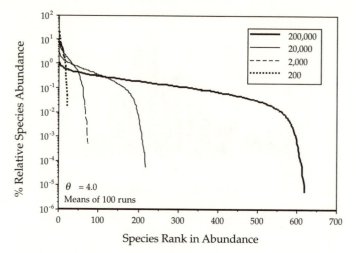

FIG. 8.17. Metacommunity dominance-diversity curves under the random fission mode of speciation, showing how changes in metacommunity size and speciation rate (not shown) affect the distribution of relative species abundance, while holding θ constant.

arbitrary abundance $< J_M$, then it matters separately what the speciation rates and metacommunity sizes are. Increasing metacommunity size while decreasing the speciation rate to hold θ constant results in a net increase in equilibrium diversity. This is because the average size of the initial populations at speciation is larger in larger metacommunities, increasing the mean lifespan of species, more than offsetting the effect of reduced speciation. The result is that more and more species are present at equilibrium between speciation and extinction. The relative abundance of the commonest species is reduced as metapopulation size increases for a fixed value of θ. Note the progressive shallowing of the curves as J_M increases in figure 8.17.

This is in marked contrast to the behavior of the metacommunity distribution under point mutation speciation. In this case, when we increase metacommunity size J_M holding θ constant, the qualitative shape of the curve remains the same

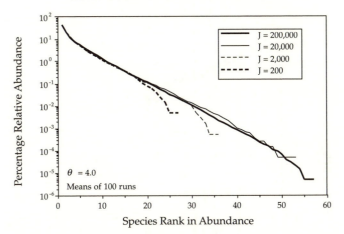

FIG. 8.18. Metacommunity dominance-diversity curves under the point mutation mode of speciation, showing that concurrent changes in metacommunity size and speciation rate that maintain a constant θ also produce the same distribution of relative species abundance. the differences at high rank are due solely to sampling effects. Compare with figure 8.17.

for the common species and species of intermediate abundance (fig. 8.18). Only the rare end of the distributions differ, and this is simply a sampling phenomenon. Species at a given rank abundance have larger population sizes in larger metacommunities, so the rank abundance at which sampling deviations from the logseries become apparent occurs at higher rank abundances in larger communities.

I have not yet had time to explore in any depth which of these distributions is a better fit to the data from diverse taxonomic groups. I offer just one example from the work of John Lynch (1989) on the genus of leopard frogs. It should be noted that the data used for this example are less than ideal because they consist of species range data as surrogates for relative species abundance, which may be problematic. Nevertheless, although Lynch argues for an allopatric origin for the majority of species in this genus of New World

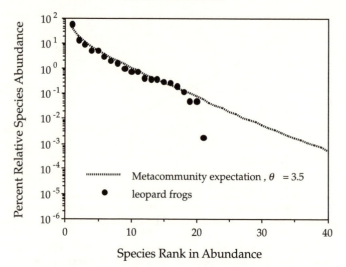

FIG. 8.19. Relationship between percent relative species abundance, as estimated approximately by relative species range sizes, and species rank in abundance for the North American genus of leopard frogs. The estimated value of θ is 3.5. Data from Lynch (1989).

frogs, the data are perhaps more consistent with point mutation speciation (fig. 8.19). If Lynch is correct, then this difference may simply be because it is difficult to distinguish a point mutation mode from speciation events that take place in small peripheral populations, as Lynch proposes for many of the leopard frog species. It is too early to predict what the preponderance of modes of speciation will be once a large sample of different taxa has been analyzed. However, in my limited sampling so far, the point mutation mode has fit more cases than the random fission mode.

The consequence of random fission speciation and a zero-sum multinomial distribution of relative species abundance in the metacommunity for the classical island-mainland problem of MacArthur and Wilson is relatively uneventful. The equilibrium distribution of relative species abundance on the island or in a local dispersal-limited community remains a zero-sum multinomial distribution. However, there is a further exaggeration of the difference between common and rare species. Rare species on the island are even less frequent than would have been expected from a logseries metacommunity distribution.

These differences raise the exciting possibility that we may be able to use the unified theory to determine what is the dominant mode of speciation in a given metacommunity. Each mode of speciation leaves a different signature in the shape of the metacommunity distribution of relative species abundance. If the metacommunity distribution is best approximated by a logseries, then we can conclude that the point mutation mode is a good approximation to the speciation process. However, if the metacommunity has a zero-sum multinomial distribution with an interior mode and a long tail of ever less frequent rare species (e.g., fig. 8.16), then we can conclude that the random fission mode is a better approximation to the prevailing mode of speciation. For example, one might argue that the distribution of total flowering plant diversity of the world (fig. 8.12) fits the random fission mode. This conclusion is still premature because the graph is based on species per family, which is an imperfect representation of the total abundance of the respective families. As I will document in the next chapter, however, there are also many metacommunity distributions that are completely consistent with the point mutation mode of speciation. The point mutation mode is more consistent with a view of biodiversity as being perfectly self-similar in its fractal geometry at all biological scales of organization.

I do not wish to leave the impression that the mode of speciation must be "either-or." The theory permits a complete continuum between the extreme of point mutation speciation, in which all species are extremely rare at origination, to the random fission mode, in which most species are abundant at origination. If point mutation and random fission are taken as representing the likely ends of a continuum, then there is a nice conjecture that comes from the neutral theory. We proved in chapter 5 that there was a complete shape continuum between the logseries and the zero-sum multinomial distribution of relative species abundance in local communities, the shape of the distribution depending upon the immigration rate. I cannot prove the following conjecture, but I predict it to be true: *There will also be a complete continuum between the logseries and the zero-sum multinomial in the statistical distributions that fit metacommunity relative abundance distributions. The shape of the fitted distribution will depend upon the relative proportion of point mutation versus random fission speciation events occurring in the particular metacommunity or phylogeny.* In putting forth this conjecture, I am not yet clear in my own mind what the analog to the immigration rate parameter m is (which generates the continuum in the case of local communities), but there must be one. This unknown but anticipated parameter will control the distribution of population sizes of species at origination.

Before leaving the subject of speciation, I return to a question initially posed in chapter 5, namely: What are reasonable rates of speciation, and are these rates consistent with the fitted values of the fundamental biodiversity number, θ? The number θ typically varies from about 0.1 to 200 or more when fitted to relative abundance data of natural communities, but commonly it is in the rate of 2 to 5 for many taxonomic groups. According to the neutral theory, under the point mutation speciation mode, for the speciation rate to generate these numbers, the speciation rate must be on the order of the inverse of twice the metacommunity size. If the speciation rate is very small, then the metacommunity size will have to be very large. Some of my colleagues have argued that the metacommunity sizes required to get a sufficiently small speciation rate are unrealistic. Or, if one sets the metacommunity sizes small enough to be realistic, then the required speciation rate becomes unrealistically high.

I don't think we know enough to reach this conclusion. How high is an unrealistic speciation rate? Recall two points from the previous section. First, the fractal nature of biodiversity appears to extend relatively homogeneously all the way down to the individual level. Suppose it is true that the events that ultimately lead to new species begin as small variations among individuals. Second, recall the serious sampling problem associated with finding all rare "species." This sampling issue is a problem in the extreme if the events ultimately leading to speciation occur, and must be detected, at the individual level. Now consider the limiting logseries distribution for the metacommunity under point mutation speciation. This mode is actually a model of infinite diversity in an infinitely large world, analogous to the infinite allele case in population genetics (Ewens 1972).

Of course, in a finite world, there will be a finite number of species (and alleles). Now consider a small sample of this world. There will be a distribution of relative species abundance in this sample. However, to fit the metacommunity logseries distribution, one assumes that the fundamental biodiversity number θ, or Fisher's α, as the case may be, would fit the data for an ever larger sample of the metacommunity, which would show an ever larger number of rare species. In a thought experiment, this could continue until every last individual were censused and classified as to whether it was or was not a new species. It might be difficult in principle to tell whether a particular individual would establish a new species, but suppose we could do so, at least retrospectively (the analog of "Lucy"). Of course, the line of descendants of most individuals is usually very short in an evolutionary sense, even those that might be considered potential founders of new species. So at this level of small-scale resolution of lineage differentiation, what would be the speciation rate? It would, according to the theory, have to be high enough to produce the observed value of the fundamental biodiversity number θ in finite samples of the metacommunity.

Now consider the sampling problem again. Most of the individual micro-variation leading to new rare species will be totally invisible to the evolutionary biologist and taxonomist. Indeed, the expectation is that by the time a species has become well enough established to be recognized and counted as a "good" species, it is already very abundant and very old. How old is old? This is still an unsolved question. For example, for many years evolutionary biologists thought that the Pleistocene was a major engine of speciation due to constant climate change, but this idea has been largely disproved by recent molecular evidence (Kling and Zink 1997). But even if the average age of most species is several million years, this does not disprove the possibility of a much faster rate operating at the level of individuals, almost all of whose descendant lineages die out rapidly. Following the argument given earlier, our sample of abundant lineages will strongly bias us *against* finding these very short-lived species, and lead to an overestimation of true species ages.

In summary, I think it is premature to dismiss the neutral theory as being inconsistent with measured speciation rates, which are potentially gross underestimates of the rate of origination events happening at the individual level, most of which will go undetected. These underestimates arise because of built-in sampling biases that automatically lead to overestimates of average lineage age.

I conclude this chapter by returning briefly to the subject of species lifespans and the average duration of taxa in the fossil record. In particular, I would like to try to reanalyze the problem of the total biodiversity of Phanerozoic marine genera that ever lived. David Raup (1991) analyzed then unpublished data of Sepkowski on the duration of 17,621 genera of Phanerozoic marine organisms in the fossil record and produced an average survivorship curve for all taxa at the generic level. Data at the generic level rather than the species level were used to overcome some of the taxonomic problems at the species level for fossil data. Raup fit these data with a proportional hazards model that came out of his neutral theory of cladogenesis, assuming constant probabilities of speciation and extinction per lineage (Raup 1978). Let $S(t)$ be the proportion of genera in the cohort of all genera that survive at least to time t, and let p and q be the rates of origination and extinction, respectively. Then the proportion of genera alive at time t should be given, according to Raup (1991), by

$$S(t) = 1 - \frac{q\left(e^{(p-q)t} - 1\right)}{pe^{(p-q)t} - q}.$$

Raup (1991) fit this equation to the observed data, and obtained estimates of genus origination and extinction rates of $p = 0.249$ and $q = 0.250$ per million years, respectively. The data and the fitted $S(t)$ function of time are shown in figure 8.20. Raup (1991) argued that the fact that the fitted speciation and extinction rates are so similar implies that diversity is nearly in equilibrium. However, recall that Raup's model does not, in fact, predict an equilibrium diversity. The only reason Raup was able to fit the proportional hazards function $S(t)$ to the data in figure 8.20 was to assume that diversity is *not*, in fact, in equilibrium. The function $S(t)$ becomes undefined when $p = q$. What is needed is a neutral theory that produces a fit to the data in figure 8.20 when the origination and extinction rates are identical.

The unified theory gives a genuine equilibrium explanation for metacommunity diversity. If we could fit Raup's data to the theory, we could get an estimate of the total number of marine genera that ever lived in the Phanerozoic. Unfortunately, we cannot truly fit the theory to the data because the size and persistence of the metacommunities that supported these marine taxa are not known. Suppose, however, we make the reasonable assumption that the lifespans of the

275

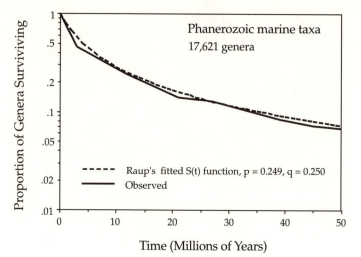

FIG. 8.20. Survivorship curve for 17,621 genera of Phanerozoic marine taxa. The dotted line is the fit of Raup's proportional hazards function, $S(t)$ for a genus origination rate of $p = 0.249$ per million years and an extinction rate $q = 0.250$ per million years. Data were compiled by Sepkowski.

genera are distributed approximately as a zero-sum multinomial. Even this assumption cannot be properly tested because we do not have data on the rarest and shortest-lived taxa. About the best we can do is to fit the data with a lognormal, which will at least give us a conservative estimate of total Phanerozoic marine diversity.

Thus, in the absence of data to fit the zero-sum multinomial distribution, we have to settle for the lognormal, which has almost exactly the same-shaped right-hand tail (for common taxa) as the zero-sum multinomial. Remarkably, the survivorship curve in figure 8.20 is almost perfectly fit by the lognormal distribution. In fitting the lognormal, I used the number of genera going extinct per million years rather than the survivorship curve. Let $S(R)$ be the number of genera going extinct R natural log units of time above or below the mode. In this case, note that I am not categorizing the number of genera dying into log base two octaves of time. Let S_0 be the modal number of genera dying. Then the equation for the lognormal after taking logs is

$$ln[S(R)] = ln(S_0) - a^2 R^2,$$

where a is a fitted constant. Since we have not observed the mode of the distribution in figure 8.20, we also need to estimate a parameter giving the position of the mode on the x-axis. We can fit the lognormal to the data in figure 8.20, and when we find the best bit by maximum likelihood, the fit is excellent (fig. 8.21). There is a slight deviation of the points for short extinction times. However, short extinction times are likely to have greater percentage measurement errors, in any case (Raup 1991), so these deviations, already small, are rendered completely unimportant. From the fitted lognormal equation, we find that the mode of the distribution is located at log_e (million years) $= -1.5$, which is equal to 223,130 years. The number of genera at the mode is $exp(8.659) = 5,762$. The value of a^2 is 0.187. The parameter a^2 is related to the variance of the lognormal by $a^2 = 1/2\sigma^2$, so the standard deviation of the lognormal in figure 8.21 is $\sigma = 1.634 log_e$ (million years).

The graph in figure 8.21 only shows the right-hand tail of the lognormal for those genera that lived long enough or were abundant enough in the metacommunity to be fossilized and discovered by paleontologists. All shorter-lived or less-abundant genera that died out in timespans shorter than a million years are not represented in the known fossil record. However, now that we have a lognormal fit to the known fossil record, we can plot the full lognormal curve, below the veil line, which in figure 8.20 is set at one million years. Figure 8.22 shows the full lognormal with its modal position at 223,130 years. The longest-lived taxa have an approximate age of $10^6 e^4$ years, or 54.6 million years. The

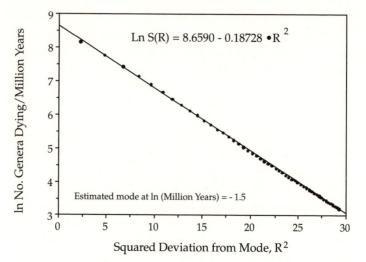

FIG. 8.21. Fit of Raup's survivorship curve to the lognormal. Only the right-hand tail of the distribution is "visible" from the fossil record, so we needed to estimate the position of the mode, which is "veiled," *sensu* Preston (1948, 1962). Maximum likelihood estimates placed the mode at 223,130 years. the modal number of genera dying is exp(8.659) = 5762. And the standard deviation of the lognormal is 1.634 ln (million years).

shortest-lived taxa are estimated to live $10^6 e^{-6}$ years, or only 2479 years.

With the full lognormal in hand, we can now estimate the total number of Phanerozoic genera that have ever lived, as well as the percentage of genera that have been found and that have never been found. From the position of the mode and the veil line, which lies 0.918 standard deviation units to the right of the mode, we can estimate the percentage of genera that lived but have no known fossil record: 82%. The remaining 18% of all genera constitute the known fossil record of Phanerozoic marine taxa.

Thus, the unified theory's estimate of the total number of genera from the Phanerozoic era that ever lived is approximately 100,000 (actual estimate is 97,894 genera). If these

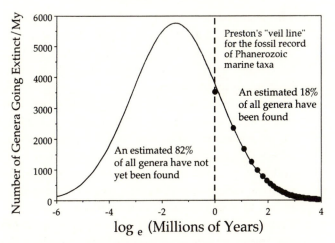

FIG. 8.22. Lognormal for the distribution of number of genera going extinct per $\log_e$ million years, fitted to the survival data of Raup (1991), from data compiled by Sepkowski. The relative position of the mode and the veil line give an estimate of 97,894 genera that ever lived in the Phanerozoic, of which 82% are estimated not yet to have been found.

numbers are approximately correct, then the fraction of all taxa that have been found is considerably higher than estimates of between 0.1% and 1%. If, however, the true distribution in figure 8.21 is an asymmetrical zero-sum multinomial, then we may have considerably underestimated the area under the left-hand tail. It is impossible to say what the shape of the left-hand tail might be until we have better relative abundance data for the known fossil taxa. There we must leave it.

SUMMARY

1. Existing neutral models of phylogeny and phylogenetic reconstruction all assume that lineages have the same probabilities of going extinct and of speciating.

2. This assumption flies in the face of strong evidence, both from ecological as well as paleobiological studies, showing that widespread and abundant lineages on average are much more resistant to extinction and have much longer lifespans.

3. Factoring in lineage abundance gives more realistic patterns of phylogeny that are qualitatively more consistent with observed patterns. Among other things, the theory predicts that most rare and endemic species are relatively recent in origin, whereas widespread abundant species are expected to be older and closer to stem ancestors.

4. The neutral theory predicts that phylogenetic clades are fractal and self-similar on all taxonomic scales from individuals on up to higher taxa. The fractal dimension of the clade is functionally related to the fundamental biodiversity number.

5. The fractal nature of clades suggests that biodiversity itself is fractal. This raises questions about the meaning of species and how to measure the speciation rate.

6. The theory predicts that the mode of speciation will leave a signature in the distribution of relative species abundance. Under point mutation speciation, the metacommunity relative abundance distribution is asymptotically logseries, whereas under random fission speciation, the distribution is zero-sum multinomial.

7. With some assumptions, the neutral theory can be used to estimate the total Phanerozoic marine diversity that ever lived; the estimate is that only about 18% of the nearly 100,000 marine genera that ever lived have been found so far.

In any case, asking how many species there are is fundamentally a *sampling* question. Total counts might seem appealing at first because they appear to have greater objectivity and precision than estimates based on sampling, but this is not necessarily so. Moreover, a total count of the world's species is not feasible—and not even approachable without heroic international investments of time, human resources, and money. Even the more modest goals of making complete species inventories of particular communities or ecosystems (e.g., "All Taxa Biological Inventories," Janzen and Hallwachs 1994) have had to face extraordinary difficulties, not the least of which is the global shortage of trained taxonomists (Wheeler and Cracraft 1997). I am not suggesting that we should abandon these efforts. Quite the contrary, systematics provides the critical information base for all of ecology and biogeography, and we need to increase the number of taxonomists. However, these efforts could be considerably strengthened by having a strong sampling strategy based on a sound theory of biodiversity. Such a theory could help answer the question of how many species there are much more rapidly, accurately, and economically.

Both Fisher et al. (1943) and Preston (1948) had sampling theories that could be used to estimate the total number of species as well as relative species abundance. Indeed, that the logseries and lognormal had sampling theories is probably the single most important reason why so much attention has been devoted to fitting these distributions to empirical data (Anscombe 1950, Bliss and Fisher 1953, Patrick et al. 1954, Preston 1962, Williams 1964, Patrick 1968, 1972, Bulmer 1974, Kempton and Taylor 1974, May 1975, Pielou 1975, Taylor et al. 1976, Slocumb et al. 1977, Engen 1978, Gray 1979, 1981, Routledge 1980, Sugihara 1980; see also reviews in Magurran 1988 and Tokeshi 1993, 1997, 1999). However, Fisher and Preston had no underlying theory of population and community dynamics from which to derive expected distributions of relative species abundance.

If a distribution fits or fails to fit the data adequately, it is unclear what significance to attach to this result because the parameters of the lognormal and logseries are generic (e.g., mean, variance, Fisher's α, etc.). These generic parameters do not reveal their derivation from fundamental processes of birth, death, migration, and speciation. This derivation is now clarified by the unified theory.

However, Preston (1948, 1962) concluded erroneously that it was possible to obtain a good estimate of the total number of species in the community from the lognormal (chapter 2). Because the lognormal is a symmetrical distribution, Preston thought that, once sample sizes were large enough to reveal the mode, one could estimate the total number of species in the community simply by doubling the number of species to the right of the mode. But as we have noted in previous chapters, virtually all observed distributions of relative species abundance for large sample sizes are negatively skewed, with a long left-hand tail and a large excess of rare species over what is predicted by the lognormal.

According to the unified neutral theory, the distribution of relative species abundance in local communities is better described by the asymmetric zero-sum multinomial distribution than by the symmetric lognormal. On larger landscapes the theory predicts that the metacommunity distribution will either be the logseries or the zero-sum multinomial, depending on the prevailing mode of speciation. Like Fisher's and Preston's theories, the zero-sum multinomial also has an associated sampling theory. Unfortunately, this sampling theory is not analytically tractable except for very small community sizes (see below). As a result, almost always the parameters of the zero-sum multinomial must be estimated by simulation. It is possible that further analytical work will enable discovery of a generating function for the zero-sum multinomial, but currently none exists. In order to make the zero-sum multinomial more user-friendly for the present, we intend to publish tables of the distribution for a very large set of possible parameter values. These will be provided on some medium such as CD-ROM or on the Internet. A fitting routine will also be available in this statistical package for fitting the user's relative abundance data.

The theory via simulation permits us to estimate the mean and variance of the total number of species in a local community, as well as the expected number of species in a subsample of the community of arbitrary size J. We can also estimate the mean and variance of the total number of species in the entire metacommunity if the size of the metacommunity is known or estimable. The total number of species in the metacommunity will have a sample variance and not be a constant because the steady-state metacommunity biodiversity is a stochastic equilibrium. Sometimes there will be more, and sometimes fewer species, depending on the actual history of random births and deaths.

The total number of species in the metacommunity can, in principle, be estimated from a finite and relatively small random sample of the metacommunity and from the density of individuals and the area occupied by the metacommunity. The estimate of the fundamental biodiversity number θ will be best if the sample truly is random. However, in most real-life situations, ecologists will have a dataset for a particular local community. This community is expected to exhibit differentiated relative abundances that reflect its size and isolation from the metacommunity. If the metacommunity relative abundance distribution is logseries, then, in principle, the fit of θ to the data should yield an estimate of the fundamental biodiversity number. As we have seen in chapter 5, the number θ is asymptotically identical to Fisher's α, which explains why the number is so stable in the face of increasing sample size. However, if the dispersal limitation affecting the local community is strong, it is likely that the true metacommunity value of θ will be underestimated. An illustration of this fitting problem can be found in the data for tree diversity in Belize forests (fig. 5.16). If all one had were the 38 1-hectare plots, the fitted value of θ would be around 11. However, when we pooled all the data from the entire length of the country, the metacommunity data fit well with a θ of 66. If one knows the value of θ for the metacommunity, then it is possible to estimate the degree of dispersal limitation and the parameter m from the degree of which local relative abundances have differentiated from the metacommunity relative abundance distribution. Because common local species are likely to be widespread metacommunity species, a better estimate can often be obtained from fitting only the common species in the local community. I will have more to say on this below.

A caveat is needed at this point. The fitting of the fundamental biodiversity number θ to relative abundance data is just that. It does not actually "test" the assumptions of the theory behind the number. To do that will require independently estimating the speciation rate ν and the metacommunity size J_M. If it is possible to obtain reliable estimates for these two numbers, then an independent calculation of θ will be possible to compare with the fitted values obtained from relative abundance data. However, as we have argued in the last chapter, all of our current estimates of speciation rate are biased low, and it is difficult to say at the present time how serious this bias may be (chapter 8).

So let us now look in more detail at the problem of finding the expected values for the relative abundance distributions predicted by the unified theory. In chapter 5 I proved that, under point mutation speciation and in the implicit spatial case, the metacommunity distribution of relative species abundance at equilibrium between speciation and extinction is completely specified by the parameter θ, a composite of metacommunity size J_M and speciation rate ν. However, in chapter 6 I showed that in explicitly spatial models, obtaining the metacommunity distribution then also requires estimating the dispersal rate m.

In principle, the exact expectations are computable for any community size, but in practice only for J up to about 10, which is far too small a community to be biologically interesting. Fortunately, the expected values can be found to an arbitrary degree of accuracy by simulation, even for very large community sizes (e.g., J_M or $J > 10^{10}$). Before discussing the simulation recipe, however, it is important to study the analytical answers for small J_M and J. Though tedious to compute, they are important checks to ensure that one's simulations are indeed generating the right expected values. For this reason, I briefly outline the procedure used in the analytical calculation of expected multispecies relative abundances. The expectations can be found exactly numerically for small J_M and J from the multispecies version of the zero-sum random walk of the ith species studied in chapter 4.

The procedure is a two-step process. The first step is to calculate the logseries expectations of relative species in the metacommunity of size J_M with fundamental biodiversity number θ from the expressions derived in chapter 5. The abundances expressed as a fraction of J_M are the expected proportional relative species abundances P_i in the metacommunity. The second step is to calculate the relative abundances in the local community. Consider a local community of size $J < J_M$, which is semi-isolated by dispersal limitation from the metacommunity ($m < 1$). Suppose that there were S species in the metacommunity of size J_M. Create a transition probability matrix whose states are all possible combinations of integer abundances of the S species that sum to the local community size J. Consider the case in which only a single death and replacement occurs. In the case of $D = 1$, transition probabilities are nonzero only between abundance states that differ by one substitution, or states that remain unchanged. Thus, the probability that species i increases by one individual and species j decreases by one individual is given simply by

$$Pr\{N_i + 1, N_j - 1, N_k, \ldots, N_s \mid N_i, N_j, N_k, \ldots N_s\}$$
$$= \frac{N_j}{J}\left[mP_i + (1 - m)\left(\frac{N_i}{J - 1}\right)\right],$$

and the probability of no change in relative species abundance is

$$Pr\{N_i, N_j, N_k, \ldots, N_s \mid N_i, N_k, \ldots, N_s\}$$
$$= \sum_{i=1}^{s} \frac{N_i}{J}\left[mP_i + (1 - m)\left(\frac{N_i - 1}{J - 1}\right)\right].$$

As J and S become larger, the number of possible integer combinations of relative abundance rapidly becomes very large. For example, for $J = 10$, the number of unique combinations of relative abundance (integers adding up to 10) is 42; but the number of corresponding combinations of 10 species for these combinations of relative abundance is 6360. For $J = 20$, these numbers increase to 627 and 4.67×10^7, respectively! Fortunately, we can reduce the dimensionality of the matrix by combining states for all species combinations that represent the same distribution of relative abundance. But to do so, we must sum all the transition probabilities for all named species for a given relative abundance distribution. A computer program can be written to do this. The necessary program calculates the transition probabilities among the states, which comprise all possible distributions of relative species abundance that sum to J. Now find the eigenvector of this numeric matrix. This eigenvector gives the unconditional (equilibrium) probability of each possible combination of relative species abundance in the local community of size J. Finally, calculate the expected abundance of the ranked species, by taking the abundance of the rth ranked species in each relative abundance distribution (state), and multiply it by the eigenvector probability for that state. Then sum these products for the rth ranked species in all states. The sum thus obtained is the analytically expected relative abundance of the rth ranked species in the local community.

A worked numerical example may make the process clearer. Suppose the metacommunity contains three species A, B, and C, each with the same metacommunity abundance ($P = 1/3$), and let the immigration probability be relatively modest ($m = 0.1$). Suppose we let local community size $J = 3$. There are then ten possible states of relative abundance and species mixtures. The ten states are: three monodominant states: (3A), (3B), and (3C); six states with two species (2A, 1B), (2A, 1C), (2B, 1A), (2B, 1C), (2C, 1A), and (2C, 1B); and one state with all three species (1A, 1B, 1C). Permutations of species order do not matter. These ten species-combination states correspond to just three relative abundance states (3), (2, 1), and (1, 1, 1). The eigenvector for the ten-state matrix is { 0.2723, 0.2723, 0.2723, 0.0292, 0.0292, 0.0292, 0.0292, 0.0292, 0.0292, 0.0079 }, corresponding to the species mixtures in the order listed above. The first three states all have the same equilibrium probability (the monodominant states) as do the middle six states (the two-species combinations) because we chose our metacommunity relative abundances of the three species to be equal. The reduced matrix with three relative abundance states has the eigenvector { 0.8169, 0.1752, 0.0079 }, corresponding to states (3), (2, 1), and (1, 1, 1).

The final step is to calculate the expected relative abundances of the ranked species. In the convention used throughout this book (and in the literature as well), species tied in rank are not given mean rank scores, but are simply assigned the tied cardinal ranks at random. This is the only way that computation of mean abundance of ranked species makes sense. In the present example, the expected abundance of the rank-1 species is $3(0.8169) + 2(0.1752) + 1(0.0079) = 2.8090$; the expected abundance of the rank-2 species is $1(0.1752) + 1(0.0079) = 0.1831$; and the expected abundance of the rank-3 species is $1(0.0079) = 0.0079$. The sum of these expected abundances is 3.0000, which is equal to J, completing the analytical result.

This numerical example is useful for demonstrating that a semi-isolated habitat or island will always differentiate a steeper dominance diversity curve than the dominance diversity curve for the metacommunity or mainland. The effect of isolation on relative species abundance is dramatized especially in this example because all the species in the metacommunity were given equal abundances.

In almost all calculations of expected values, the preceding analytical approach will not be pursued because of the computational problems posed by large J_M and J. I now discuss a simulation algorithm which works for large and small J. The following recipe is due to Warren Ewens (pers. comm.). Call the quantity $\theta/(\theta + j - 1)$ the *species generator*. It represents the probability that the next individual sampled is of a new species not previously collected. Draw the first individual, and label it species 1. Now collect the second individual, and draw a random uniform number x between zero and one. If x is less than the value of the species generator for $J = 2$, i.e., $x < \theta/(\theta + 1)$, then individual 2 is a new species. However, if $x > \theta/(\theta + 1)$, then the second individual is another specimen of the first species. Now collect the third individual and draw a new random uniform number x between zero and one. If x is less than the species generator for $J = 3$, i.e., if $x < \theta/(\theta + 2)$, then the third individual is a new species; otherwise, it belongs to a species previously encountered.

In this case, since the individual is of a previously encountered species, we must determine to which of these already collected species the individual belongs. Do the following. First, calculate the current fractional abundance of each species from the first species to the last species found. Suppose the current abundance of the ith species is n_i before adding the jth individual. Then the fractional abundance of the ith species is $n_i/(j-1)$ before the jth individual has been classified. Add these current fractional abundances in any order from the first species to the last, and record the partial sums (yielding the cumulative distribution). Now draw a random uniform variable x from zero to one and determine in which interval of the cumulative distribution of current relative abundances the random number lies. Add one to the abundance of the corresponding species. Now collect individual $j+1$, and repeat the steps above to determine if it is a new species or an individual of a previously collected species. Repeat this process until the sample size J has been reached. A single run of this algorithm produces one sample distribution. To estimate the sample mean and variance of metacommunity relative species abundance to an arbitrary degree of accuracy, run this algorithm an arbitrarily large number of times, and compute the ensemble mean and variance of the abundance of the rth ranked species for a sample of fixed size J. In my experience, there is little change in the estimates of ensemble mean and variance after 100 runs. The logic of the algorithm is depicted as a flow chart in figure 9.1.

The previous algorithm is appropriate for finding the expected metacommunity distribution under no dispersal limitation ($m = 1$). However, if one wants to compute the expected relative abundance distributions for local communities or islands that are subject to dispersal limitation, another approach is needed. There are several ways to compute the expected distribution under dispersal limitation. One way is approximate and works for relatively small sample sizes. Since I am assuming for the moment that θ and m are known, we can modify the species generator in the algorithm above to include dispersal limitation: $\theta \cdot j^{-\omega}/(\theta+j-1)$, where $m(j) = j^{-\omega}$. Caution is necessary in using this generator, because it is only approximate; recall that there is as yet no analytical expression for the species-individual curve

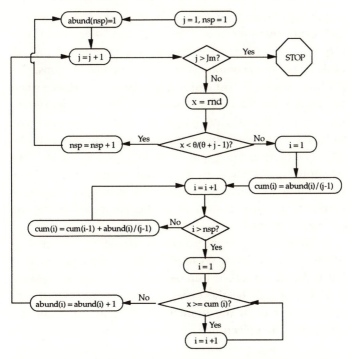

FIG. 9.1. Flow chart of the logic of the algorithm for computing the metacommunity distribution of relative species abundance for known θ and given sample size J_M. Variable *nsp* is the current number of species in the sample up to $j-1$ individuals. Variable *abund* is the vector storing the current abundances of the species collected so far, and variable *cum* is a vector representing the cumulative distribution of the fractional relative species abundances of the species collected so far. Function *rnd* delivers a random variable distributed uniformly between zero and one, and it gives a new pseudorandom value of the variable every time it is referenced.

under dispersal limitation. A more laborious but also more accurate method is to simulate the island-mainland coupled dynamical equations. The mainland or metacommunity distribution is found by the algorithm given above for a given θ and a metacommunity sample much larger than the size of the local community. Then ecological drift in the local community on an island is simulated for a given value of m and local community size J. A set of simulation runs is performed, and, as before, the mean and variance of the rth ranked species is computed over the ensemble of runs.

I turn now to the issue of fitting relative abundance when the parameters θ and m are unknown. For fitting purposes, the value of J is already known and equal to sample size, so fitting the relative abundance distribution reduces to the problem of estimating the most likely values of the two remaining parameters, θ and m, given the observed relative abundance data. The expected distribution of relative species abundance is then fit to the observed data by the method of maximum likelihood.

In theory, the general procedure is as follows: Let $\Psi = \{\theta^*, m^*\}$ be a particular set of trial values for the parameters. Let n_r be the observed number of species with abundance r, and let $E_r\{\Psi\}$ be the expected number of species with abundance r from the unified theory. The log likelihood function is then given by

$$F(\Psi) = \sum_{r=1}^{J}\left\{ n_r \log\left[\frac{E_r(\Psi)}{n_r}\right] - [E(\Psi) - n_r] \right\}.$$

We then find the maximum of $F(\Psi)$ as we vary Ψ. In general, relative abundance data are sparsely distributed over all possible abundances $< J$, so it is often more convenient to consider the expectations for grouped observations, such as in doubling classes of abundance in the case of Preston-type plots.

Unfortunately, it is only possible to write down the analytical expectations in $F(\Psi)$ for the metacommunity case, not the local community case, and these are only practically computable for small J_M. Therefore, an alternative method is required, which is to fit θ and m sequentially. The sequential fitting method works well because the estimate $\hat{\theta}$ is very insensitive to variation in m. Recall from chapters 4 and 5 that the locally common species on average are also the most common species in the metacommunity. Because of the stability of the law of large numbers, the local relative abundances of common species are also expected to be very close to their metacommunity relative abundances (on a percentage basis). Rare species tend to be relatively rarer in the local community than in the metacommunity (chapter 5). Therefore, in step 1, only the commonest species in the local community are used to fit the metacommunity parameter θ. For step 1, the expected metacommunity relative abundances for a random sample of size J are obtained from the algorithm given in figure 9.1. As a convenient rule of thumb, use all species more common than the median abundance to fit θ in step 1 of the estimation procedure. Once θ has been estimated, $\hat{\theta}$, then one proceeds to step 2 to estimate m from simulations to obtain the expected distribution of all the species, common and rare. The value of θ is fixed at $\hat{\theta}$, and local community dynamics are simulated for various values of m. The final estimate of m, $\hat{m}$, is that value which yields the maximum likelihood. This sequential estimation procedure was used to obtain θ and m for the BCI and Pasoh 50 ha plots (see figs. 5.8 and 5.9), as for the other communities discussed later in this chapter.

It is worth emphasizing that there are actually two quite different sources of variance in empirically derived distributions of relative species abundance in local communities. The first is simple sampling variance. Assuming that there exists a "true" distribution that describes the relative abundance of all species currently present in the community, then subsamples of the total community will have an associated sampling variance. This variance is the only component of variance expected from a static sampling process (e.g., the logseries or the lognormal). One can estimate the sampling variance for a dominance-diversity curve by computing the expected cumulative distribution function of the relative species abundances for the fitted values $\hat{\theta}$ and $\hat{m}$, and then randomly sampling this distribution with a random uniform variable until the desired sample size is reached. If sample sizes are a substantial fraction of the community being sampled, then the distribution function should be created from species abundance counts (i.e., not be fractional relative species abundances), and the sampling should be done without replacement. The sampling variance is then estimated over a large number of independent trials of this random-draw procedure.

The second component of variance is perhaps the more important, and results from the dynamical process of demographic stochasticity that underlies ecological drift in the unified theory. This variance is estimated by simulating ecological drift in the local community for the fitted parameters $\hat{\theta}$ and $\hat{m}$. One can then compute the variance in the abundance of the ranked species over an ensemble of simulation runs. In these simulation runs there is no sampling variance because all the species abundances are completely and exactly known. The total variance observed when one is sampling a real community is thus the sum of the sampling variance and the variance due to demographic stochasticity. The total variance is the measure of variation that should be used in designing statistical tests of the fit of the distributions predicted by the unified theory.

I now briefly discuss hypothesis testing about relative species abundance distributions. Typical questions likely to be asked about relative abundance include the following: Is a particular observed distribution consistent with the hypothesis of ecological drift? Are the sample distributions from two different communities statistically different? Is there excess dominance in the community over what drift alone can explain? Consider a temperate tree community example. In her book on the forests of eastern North America, Braun (1950) published numerous inventories of relative tree species abundance in many stands. Consider two mixed mesophytic forests on the Cumberland Plateau of southern Kentucky, one in Bell County and one in nearby Perry County. In the Bell County forest, the rank-1 species was sugar maple, and the second most abundant species was basswood. However, in the Perry County forest, the most abundant species was tulip poplar, and the second most abundant was beech. Based on the aggregate of a larger number of samples of stands over the Cumberland Plateau of Kentucky, I estimated the fundamental biodiversity number for the mixed mesophytic forest metacommunity. Recall that aggregating small samples from all over the metacommunity tends to overcome the effects of dispersal limitation, and the resulting distribution is expected to be closer to a logseries (chapter 6). This distribution is shown later (fig. 9.6).

The dominance-diversity curves of the Bell County and Perry County forests are shown in figure 9.2 along with the expected metacommunity distribution for $\theta = 5.2$ and $m = 0.2$, fitted by the methods given earlier. For the simulations I assumed an annual mortality rate of one percent. Based solely on the dominance-diversity curves by themselves, one can conclude that these two forests are consistent with drift and the zero-sum multinomial distribution. However, a closer examination at the species level shows that, although their dominance-diversity curves are consistent with zero-sum dynamics, these two forests are also significantly different in species composition. Figure 9.3 shows the expected distribution of relative abundance for sugar maple, beech, tulip poplar, and basswoood in terms of the percentage relative abundance of the species in the metacommunity. Beech was completely absent from the Bell County forest (panel B). The Perry Co. site also had too little sugar maple (panel A) and too much tulip poplar (panel C) to be neutrally dispersal assembled from the metacommunity. These conclusions of course depend on the assumption that the metacommunity composition of the mixed mesophytic forest has been

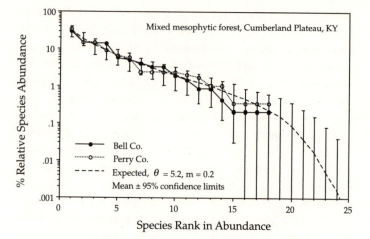

Fig. 9.2. Relative species abundance distributions for two temperate tree communities in Kentucky. The two stands were of mixed meso-phytic forest on north-facing slopes a few miles apart on the Cumberland Plateau. The fitted parameter values θ and m yielded confidence intervals that completely bracketed each distribution.

measured accurately. A far larger regional sample would be preferable to check this assumption.

Before leaving the topic of hypothesis testing, I wish to illustrate the dangers of judging distributions of relative species abundance by visual inspection alone rather than by statistical test. Individual "snapshots" of relative species abundance distributions taken at single points in time and space may look very different from lognormals—that is, zero-sum multinomials—even when the underlying process is indeed a stochastic, zero-sum game. Depending upon the parameter values and the disturbance rate to which a community is subjected, the variance in relative abundance distributions over time can be large or small under zero-sum ecological drift. Figure 9.4 shows Preston-type plot of relative species abundance for sixteen simulations of a model community undergoing zero-sum drift. The local

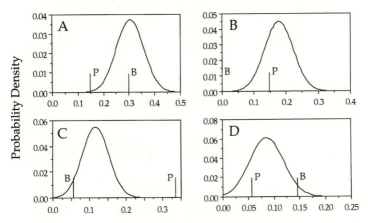

Percent Relative Species Abundance

FIG. 9.3. Relative abundance distributions for selected individual species in 2 stands on north-facing slopes in mixed mesophytic forests in Kentucky. The curves represent the distribution of the expected percentage relative abundance of a given species based on the meta-community relative abundances of each species, and the vertical lines are the observed percentages in the three stands. The letter *B* indicates the Bell Country forest, and the letter *P* the Perry Country forest. (A) Sugar maple, (B) beech, (C) tulip poplar, (D) basswood.

community size was $J = 1600$ individuals, and the meta-community had a θ of 50. The immigration rate m was 0.01, and the death rate per disturbance cycle was 400. In each run the distribution of relative species abundance was initialized as a random sample of the metacommunity. The stochastic dynamics of the community were then simulated for one thousand disturbance cycles before the relative abundance distribution was calculated. This represents 250 complete turnovers in the local community, more than adequate time for it to achieve stochastic equilibrium. The top three rows of figure 9.4 are snapshot distributions of relative species abundance for each of the first twelve simulations. The bottom row of distributions represents four of the more extreme cases that were encountered in one hundred runs.

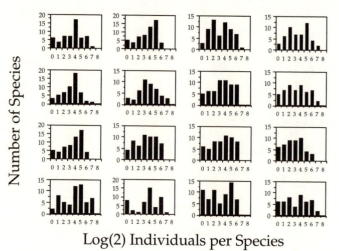

Log(2) Individuals per Species

(y-axis label: Number of Species)

Fɪɢ. 9.4. Stochastic variation in the relative species abundance distributions of a model community undergoing zero-sum ecological drift. The metacommunity value of θ is 50. The local community size J is 1600. Immigration rate is 0.01. One quarter of the individuals are replaced per disturbance cycle. Relative abundance distributions were calculated after 1000 disturbance cycles (after 250 turnovers of the community). The top twelve panels were the results of the first twelve runs. The bottom four panels are four extreme cases found among the one hundred runs.

As is clear from the figure, the local community stochastic equilibrium can exhibit considerable variance, especially if the local community is subject to high disturbance rates (high turnover rates), as in the case illustrated. Many of the individual snapshot distributions do not look very much like a zero-sum multinomial, yet the underlying community dynamics do generate the expected stochastic equilibrium, zero-sum multinomial relative species abundance distribution, as shown in figure 9.5.

In a symposium on benthic marine ecology, Lambshead and Platt (1985) published a review and general critique of Preston's hypothesis. They argued that the evidence for the pervasiveness of lognormal distributions, not only in

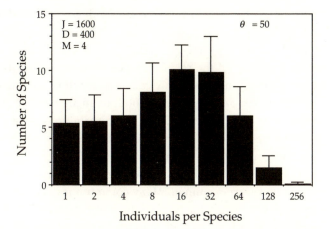

Fig. 9.5. Stochastic equilibrium relative species abundance distribution for the model community studied in figure 9.4. The means of one hundred independent simulations are shown. Error bars represent ±1 standard deviation. The magnitude of the error bars depends particularly on the disturbance rate ($D > 1$ or multiple deaths per disturbance cycle; see chapter 4).

benthic communities, but in terrestrial communities as well, was not compelling, contrary to claims in the literature (e.g., May 1975, Sugihara 1980). However, they made no formal test of goodness of fit to the lognormal. Lambshead and Platt (1985) qualitatively compared the shapes of a large number of Preston-type plots of relative species abundance distributions from various sources and concluded by visual inspection that they were rarely lognormal; or, if they were, that the lognormals were only obtained in the case of aggregated heterogeneous samples.

Lambshead and Platt did not have the benefit of the unified theory in making their argument. Their failure to perform goodness of fit tests is problematic, but such tests may still fail without a dynamical hypothesis to explain the relative abundance distributions. That is, the observed variation about the expected distribution may be larger than

predicted by the sample variance alone, due to the underlying stochastic demographic process. The conclusion from
figures 9.4 and 9.5 is that one cannot merely visually inspect
snapshot samples of the relative species abundance distribution in a community and safely conclude much of anything.
The differences among the distributions in figures 9.4 are
considerable despite the fact that the distributions were all
produced by the identical dynamical process in a community undergoing zero-sum ecological drift. This is not to say
that Lambshead and Platt were wrong about many of the
communities they examined. Many communities may fail to
obey zero-sum dynamics because they are so severely and
frequently disturbed, that all limiting resources are not, in
fact, consumed, or because the assumptions of a zero-sum
game are not met.

Sampling issues and hypothesis testing regarding relative
species abundance distributions are important because these
distributions are widely used in attempts to measure the
"health" of ecosystems (Gray 1981, Kevan et al. 1997). The
rationale for these tests arises from the usually untested
assumption that if communities exhibit lognormal-like distributions of relative species abundance, then they are health-
ier than if they do not. The unified theory provides some
justification of this assumption. Relative abundance distributions are expected to deviate from the zero-sum multinomial
if the disturbance rate is so severe that limiting resources are
not totally saturated, so that zero-sum dynamics do not apply
to the community (chapter 3). However, some researchers
have concluded that finding non-lognormal-like distributions means that the communities are unhealthy. This is
not necessarily so. For example, geometric-like distributions
may be steady-state zero-sum multinomials in communities that are species poor simply for reasons of moderate
to extreme isolation from the source-area metacommunity.
The critical issue is whether the distribution remains a

zero-sum multinomial, not whether it is a geometric-like or a lognormal-like multinomial.

For example, Kevan et al. (1997) compared the pollinator communities over an eight-year period in thirteen blueberry heaths in New Brunswick, Canada. Some of these heaths and the surrounding forest had been treated long-term with pesticides, whereas others were nearly pesticide free. Kevan et al. reported that they found lognormal distributions of pollinator species abundance in the pesticide-free heaths, but often did not in the pesticide-exposed heaths. They concluded that the lognormal was a reliable and sensitive diagnostic tool to diagnose ecosystem health. This may be so, but at least some of the non-lognormal distributions that Kevan et al. report are fit quite well by the zero-sum multinomial distribution.

Relative species abundance distributions have also been used to assess ecosystem health by aquatic ecologists studying benthic communities (Patrick 1968, Gray and Mirza 1979, Gray 1981, 1983). Initially it was thought that, if the lognormal was the expected distribution for undisturbed communities (Preston 1980), then departures from the lognormal, particularly toward geometric-like distributions of relative species abundance with higher dominance, could be used as sensitive indicators of disturbance, particularly by pollution (Gray and Mirza 1979, Andrews and Rickard 1980, Gray 1981, Mirza and Gray 1981, Bonsdorff and Koivisto 1982, Thompson and Shin 1983) and other environmental stresses (Gulliksen et al. 1980, Hicks 1980, Ortner et al. 1982, Soulsby et al. 1982).

As time passed and more communities were sampled, however, ecologists discovered that many unpolluted benthic communities exhibited geometric-like, not lognormal-like distributions. In these studies, most of the sample sizes were fairly small, so Hartnol et al. (1985) carried out a study to see whether relative abundance distributions would become more lognormal-like as sample sizes were increased. They

301

took very large samples of the benthic fauna of unpolluted subtidal and tidal sands along the coast of the Isle of Man. To their surprise, even in these unpolluted sites and in large samples, they obtained geometric-like, not lognormal-like, distributions of relative species abundance. Hartnol et al. therefore concluded that departures from the lognormal could not be reliably used as an indicator of pollution disturbance or stress.

From the perspective of the unified theory, we need to qualify their conclusions. As we have seen, the theory predicts that if a local community is subject to strong dispersal limitation, then this community will exhibit a steady-state dominance-diversity curve that is geometric-like—even in the absence of any exogenous stresses such as pollution. Clearly, a stronger case for a pollution effect could be made if an actual *change* were observed in a given community from a lognormal-like to a geometric-like zero-sum multinomial, and if this change were associated with an alteration in pollution level. Such a change would be predicted by the theory if pollution were to cause local extinctions of some species in the community. Then, under the zero-sum rule, the surviving species in the species-poorer community would exhibit a steeper and more geometric-like dominance-diversity curve, with higher dominance of the common species. In many communities, time-series data are unfortunately not available, so whether the community has lost species, or whether there has been an increase in apparent dominance, is generally not known. However, even for single snapshot samples, it is possible to test whether the relative abundance distribution has excess dominance over that predicted by the unified theory. I will illustrate a test showing excess dominance in a tropical tree community in chapter 10.

How general is the unified theory? Until this point I have illustrated the predictive power of the theory with examples mainly drawn from closed-canopy tree communities—those with which I have the most familiarity. How well does the theory fit the data for communities of organisms that are very different ecologically and trophically from trees? In the first figure of this book, I illustrated the dominance-diversity curves of a wide variety of communities (fig. 1.1). Any durable theory of biodiversity and biogeography must successfully describe and predict dominance-diversity curves from the first principles of a dynamical theory of communities. In what follows, I merely seek to demonstrate that the unified theory does a remarkably good job of fitting many dominance-diversity distributions. However, the theory by no means fits all distributions, and therein perhaps lies the greater biological interest. I examine some failures to fit at the end of this chapter.

My first example is the fit of the theory to the relative abundance distribution of mixed mesophytic forest on the Cumberland Plateau of Kentucky (fig. 9.6). This distribution is based upon stands scattered over an area of approximately 13,000 square miles. The regionally pooled data, as expected, are fit quite well by the metacommunity distribution, assuming no dispersal limitation ($m = 1.0$). The fitted value of the fundamental biodiversity number θ was 5.2.

Another metacommunity distribution that is fit quite well is the set of all samples of the planktonic copepod community of the northeastern Pacific gyre (McGowan and Walker 1993) (fig. 9.7). This open-ocean copepod community is likely to be reasonably thoroughly mixed even without the pooling of plankton samples, but once again there is little hint of dispersal limitation. The estimated value of the fundamental biodiversity number θ in this community was 32, and the estimate of the immigration rate was 0.9, very close to unity.

The next example is of a tropical bat community. Charles Handley and coworkers censused bats on and near Barro Colorado Island for 63 months during the decade from 1975 to 1985. Kalko et al. (1996) recently summarized these

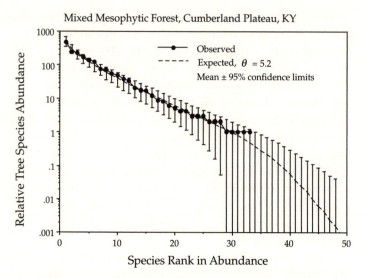

FIG. 9.6. Fit of the metacommunity distribution to pooled data from stands of mixed mesophytic forest on the Cumberland Plateau, Kentucky. The estimated value of the fundamental biodiversity number θ is 5.2.

long-term studies and published dominance-diversity curves for the BCI bat community for the first time. Bats are highly mobile organisms, and one might expect that their relative abundances would therefore conform more closely to the metacommunity logseries expectation, and this seems to be the case (fig. 9.8). However, the fit to their total BCI bat community is not particularly good. Kalko et al. discuss the role of dispersal versus niche in assembling the BCI bat community, and they note that the bats differ not only in their trophic guild, but also differ in modes of foraging within these trophic guilds. There are some deviations about the expected line that are larger than those seen in earlier examples; and these deviations may reflect departures from complete zero-sum dynamics in the BCI bat community. Nevertheless, the distribution is a fair fit to a zero-sum multinomial with no dispersal limitation. The estimated

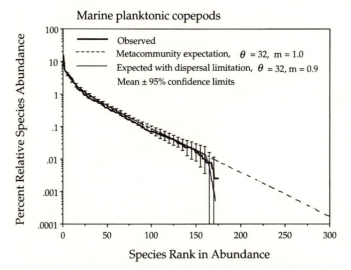

Fig. 9.7. Fit of the metacommunity distribution to the planktonic copepod community of the northeastern Pacific gyre, without dispersal limitation ($m = 1.0$) (diagonal dashed line extending off to the right), and with minor dispersal limitation ($m = 0.9$). Data from McGowan and Walker (1993).

value of the fundamental biodiversity number for the BCI bat community θ is 6.0.

I turn now to an example of an insect community—a desert bee community in Utah. One of the most remarkable bee communities in the world is the very rich desert bee community of the American Southwest. The general view has been that many of the bee species are specialists on one kind of plant, such as one species of desert annual. More recent evidence suggests that these bees may be much more opportunistic and facultative in their foraging (Tepedino, pers. comm.). This seems reasonable given the notoriously variable annual plant community of the desert Southwest, and the dominance-diversity curve for solitary bees of the family Andrenidae supports this conclusion (fig. 9.9). Note that the dominance-diversity curve falls away

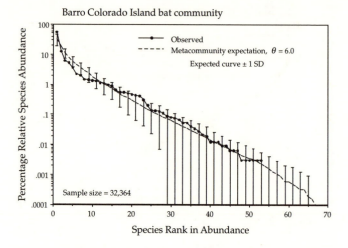

FIG. 9.8. Dominance-diversity curve for the bat community on Barro Colorado Island, Panama, and the fit of the metacommunity distribution of the unified theory, for $\theta = 6.0$. Note that there is no evidence of dispersal limitation in these data. Part of the reason may be that these abundance represent capture records through time, so that there is a time factor that may have helped overcome the dispersal limitation of some species. Data are from Kalko et al. (1996).

from the metacommunity curve for rare species. In the unified theory, this is indicative of isolation and dispersal limitation. The estimate of m for this bee community is 0.2.

A community which shows evidence of extreme isolation and dispersal limitation is the freshwater fish community of Caño Maracá in the headwater tributaries of the Rio Negro in southern Venezuela (Winemiller 1996). Very large fish collections over a number of years have produced a detailed record of relative species abundances, even for very rare species (fig. 9.10). The steep falloff of the dominance-diversity curve from the metacommunity curve begins at about fifty species. The rarest 40% of the species collectively constitute just 1.95% of all individuals. The value of θ estimated from the common species is 25. The estimated

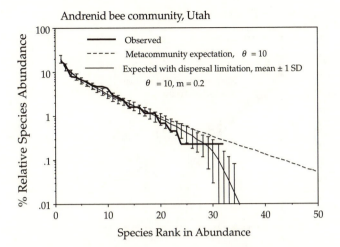

FIG. 9.9. Desert bee community, family Andrenidae, in Utah. The community has an estimated θ of 10 and also shows evidence of dispersal limitation ($m = 0.2$). Data courtesy of Vince Tepedino of Utah State University and the USDA Apis Research Service.

probability of immigration, m, is very small—only one in 10,000 births for this headwaters fish community.

It might not be surprising that a tropical river headwaters fish community is so isolated, but what about a bird community? Richard Holmes has been studying the long-term dynamics of the bird community in the forests of Hubbard Brook, New Hampshire, for more than 30 years (Holmes et al. 1986) (fig. 9.11). Many of the Hubbard Brook bird species are summer breeding residents and winter migrants. I examined the relative abundances of the subset of species that are primarily insectivores, assuming that these species would represent a guild that might reasonably be expected to obey zero-sum dynamics. The data are very well fit by a zero-sum multinomial with $\theta = 5.0$ and $m = 0.2$. This suggests that approximately one out of every five insectivorous birds in the community is an immigrant and, conversely,

307

Caño Maracá fish community, Venezuela

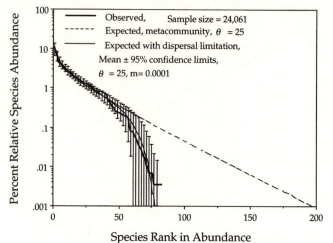

Fig. 9.10. Tropical freshwater fish community in a system of tributaries in the upper watershed of the Rio Negro. Note the extreme degree of isolation evident for this fish community: the estimate of the probability of immigration per birth m is only 0.0001. The estimated fundamental biodiversity number θ is 25. Data courtesy of K. Winemiller.

that four out of five birds were born locally in the Hubbard Brook forest.

The final example is the dominance-diversity curve for the entire avifauna of Great Britain (Gibbons et al. 1993), which I presented as a Preston-type plot in chapter 2 to illustrate the shape of the zero-sum multinomial (fig. 2.6). When plotted as a dominance-diversity curve, we observe the now familiar phenomenon of the departure of the observed distribution from the metacommunity expectation for rare and very rare species (fig. 9.12). I show this distribution primarily to make the point that dispersal limitation can be detected on arbitrarily large spatial scales, as for example, on the scale of Great Britain. There are many problems with the precision of estimating bird population sizes in such a large area (Gibbons et al. 1993). However, in the well-studied British

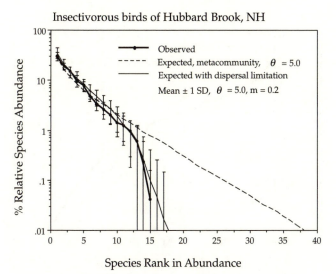

FIG. 9.11. The long-term average dominance-diversity curve for 15 species of insectivorous birds in the summer breeding bird community of Hubbard Brook, New Hampshire, USA. The estimated value of the fundamental biodiversity number θ is 5.2, and the estimated per capita probability of immigration m is 0.2. Data from Holmes et al. (1986).

avifauna, it is almost certainly the case that the absolute abundances of the rarest species are better known than the absolute abundances of the commonest species. This means that we can put considerable confidence in the shape of the dominance-diversity curve at the rare end of the distribution. The common end is also well determined for purposes of a dominance-diversity plot because relatively large absolute errors in estimating the abundances of common species become insignificant on a log scale (fig. 9.12). It is possible to fit the metacommunity distribution to these data, which yields a value of the fundamental biodiversity number θ of 25. However, I was not able to fit the entire community of 37.9 million breeding pairs of birds for the dispersal limitation parameter m because the community size is too large for my computer.

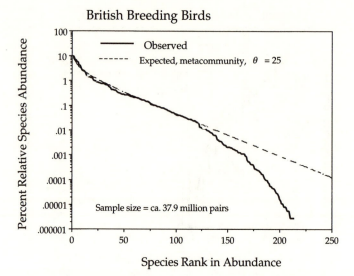

FIG. 9.12. The dominance-diversity curve for the entire British breed-ing avifauna. The estimated fundamental biodiversity number θ is 25. Data from Gibbons et al. (1993).

I conclude this chapter with a discussion of the robust-ness of the unified theory of biodiversity and biogeography. By *robustness* I refer to the ability of the theory to with-stand violations of its assumptions and still be predictively useful. A case in point is the excellent fit of the theory to the relative species abundance distribution of the entire British breeding avifauna, given just above (fig. 9.12). The fit to the British avifauna, which includes everything from seabirds to raptors to warblers, suggests that, at least on such macroscopic scales, the differences in ecology among the species become obscured by biogeographic factors operat-ing on very different spatial and temporal scales. Moreover, differences in ecology may tend to be overwhelmed by the absolutely vast differences in relative species abundance rep-resented by the distribution, which ranges over seven orders of magnitude, from species having a single breeding pair to species with tens of millions of breeding pairs. Such a good fit may also be an exception, however. It is too early to say. Certainly there is abundant and growing evidence that energetic and biomechanical scaling laws exist that dictate that the body size of species will be strongly and negatively correlated with landscape-level species abundance in many taxa (e.g., Brown 1995). At this early stage in the theory's development, it is impossible to give definitive answers to questions about its robustness. Nevertheless, there are still some useful things that can be said about robustness at this point.

First, we can dispense with the concern that the theory may "fit everything"—a concern raised, for example, by the excellent fit to the entire British breeding avifauna. A neutral theory that fitted everything would not be very useful because it would fail to instruct us about the assembly of natural communities. Exactly when and how a good formal neutral theory fails should be as interesting, if not more so, as when and how it succeeds.

Fortunately, it is not difficult to find communities in which the dominance-diversity curve predicted by the unified theory does not, in fact, fit the data. For example, it fails to fit the relative abundance data for all birds censused in a 10 ha plot of forest in Manu National Park, Amazonian Peru (Terborgh et al. 1990). Like the British avifauna, this survey included all taxa, irrespective of ecology, from hummingbirds to guans. But unlike the British avifauna, the distribution of Manu birds deviates significantly from the best-fit theoretical metacommunity distribution. The Manu sample was of a very local bird community, in which the range in relative abundance was much smaller (only two orders of magnitude) than in the example of all breeding birds of Great Britain. Whatever the cause of differences in fit, the theory does not fit the observed distribution for Manu (fig. 9.13).

Attempts to fit the metacommunity distribution to the common species failed in two interesting ways. If the

311

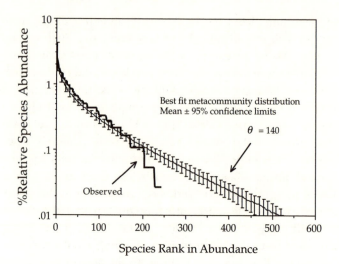

FIG. 9.13. The observed dominance-diversity curve (heavy line) for all species of bird censused in a 10 ha sample of mature forest in Manu National Park, Peru. The best fit metacommunity distribution yields a θ of 140. However, the metacommunity distribution fails to predict the high abundance of the midrank species. No attempt was made to fit the probability of immigration, m in this case, although it is clear that dispersal limitation does affect the curve for species that are rarer than rank 150. Data from Terborgh et al. (1990).

fundamental biodiversity number is fit to the fifty most abundant species, then there is an excess of abundance in the midrank species over expected abundances. On the other hand, if the midrank species are also included in the set of species fit by the metacommunity distribution (the 150 most abundant species), then the most abundant species display excess dominance. That is, the observed abundance for the most abundant species is significantly greater than the predicted abundance of these species by the metacommunity distribution. We can begin to explore some reasons for the lack of fit in this case. A major violation of the assumption of zero-sum dynamics is a likely candidate. Terborgh et al. (1990) also presented data on the average masses of the different species in the Manu forest. It is interesting that there is only a weak negative and nonsignificant correlation between abundance and body mass in these data (R^2 = 0.096) (fig. 9.14). Among other things, this lack of correlation may reflect the fact that the spatial scale of resource dependence that determines bird abundances is very different among different trophic guilds, so that the 10 ha spatial scale of reference is inappropriate for all species. For example, raptors exploit prey resources over a much larger area than 10 ha, whereas many small insectivores might spend their entire lives in the 10 ha plot. This underscores the fact that *one of the major challenges in testing the unified theory is determining the appropriate spatial and temporal scales for defining both local communities and the metacommunity.* This may be a far easier task in general for communities of sessile organisms.

The fact that biomass and abundance are largely decoupled in this bird community is further supported by the following observation. The distribution of relative species abundance is negatively skewed, as predicted by the zero-sum multinomial (fig. 9.15A). However, the distribution of body mass is positively skewed even as a log-transformed Preston-type plot (fig. 9.15B). The total community biomass distribution can be computed by plotting the distribution of the *product* of mean body mass times the species abundance per 10 ha. This distribution of the biomass of total bird species populations in the 10 ha Manu plot is nearly perfectly lognormal in shape (fig. 9.16). This finding suggests that there is little reason to expect a strong relationship between body size and abundance in local communities where the resources supporting that community may or may not be completely local. The whole issue of body size has not been treated at all thus far by the unified theory. The linkages between theories of body size scaling rules, zero-sum dynamics, and the unified theory are likely to be very fertile areas for future theoretical exploration, but this lies beyond the scope of the present book.

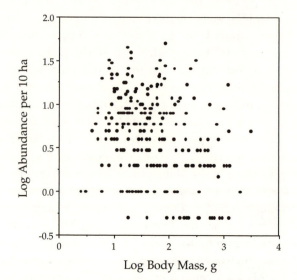

FIG. 9.14. Lack of relationship between average body mass of bird species in a 10 ha plot in Manu National Park, Peru, and the abundance of the species in the plot. Note that a few species have estimated abundances in the 10 ha plot that are less than unity. These were species that may not have breeding pairs in the plot. Data from Terborgh et al. (1990).

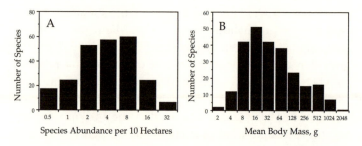

FIG. 9.15. Distributions of bird species abundance (A) and of mean body mass in grams (B) for the bird community in the 10 ha plot in Manu National Park, Peru. Data from Terborgh et al. (1990).

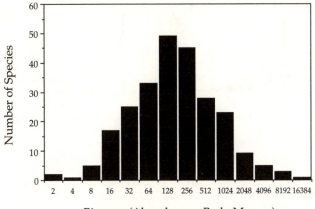

Biomass (Abundance x Body Mass, g)

FIG. 9.16. Distribution of total bird species population biomass in a bird community in Manu National Park, Peru, expressed as abundance per 10 ha times mean body mass in grams. Data from Terborgh et al. (1990).

I now illustrate one more case in which there is a poor fit of the unified theory. These are data on rain-forest canopy Coleoptera collected and identified by Nigel Stork and colleagues from the canopies of ten trees in Borneo (Stork 1997). These samples, and others on chalcid wasps not shown, are characterized by an enormous fraction of singleton species. In the sample illustrated in figure 9.17, 499 out of 859 species (58%) were collected only once. At the other extreme of abundance, the top ten most abundant species constitute a third of all individuals (32.2%). The unified theory in its present form fails to fit these data. If we use the top ten species to estimate the fundamental biodiversity number, the resulting metacommunity curve is far too steep for the observed dominance-diversity curve, and the fitted θ considerably underestimates the species richness observed in the sample. One explanation for this failure might be that a very large fraction of the singletons represent nonresident species that were temporarily resting on the sampled trees

315

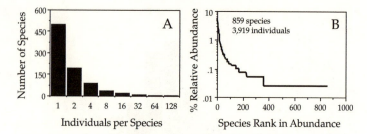

FIG. 9.17. Relative abundance distributions for Coleoptera collected from 10 trees in a rainforest in Borneo. (A) A Preston-type plot showing the enormous number of species collected only once. (B) The corresponding dominance curve. There were 859 species found among 3919 individuals. Data from Stork (1997).

at the time of the collection. Stork (1997) states that many of the extremely rare species probably do not feed on the tree species from which they were collected. There are no host plant records to support this speculation, but it is a reasonable hypothesis. Most of the beetle species are extremely small, and they are likely to be capable of long-distance dispersal, carried by the wind. The data in figure 9.17B are exactly what one would expect if the dispersal kernels of these beetles were highly leptokurtic with extremely long "fat" tails, extending far away from individual host trees.

It should be noted that that the data in figure 9.17 are not intrinsically incompatible with neutral theory. In a more refined, spatially explicit version of the theory, we could build in a strongly leptokurtic dispersal kernel that would have the effect of increasing the proportion of rare, long-distance dispersers in local communities. This dispersal function could apply equally to all species, and the theory would still remain completely neutral, according to the definition of neutrality given in chapter 1.

The key assumptions of the unified neutral theory of biodiversity and biogeography really boil down to only two. The first is that all species are identical on a per capita basis in their probabilities of birth, death, and dispersal. The second is that these species are locked in a life-or-death zero-sum game competing for the same, shared, limiting resources. How robust is the unified theory to violation of these assumptions? Ironically, if one believes that most ecological communities are fundamentally niche assembled, then the answer must be "very robust," because the theory does a remarkably good job of describing local and landscape patterns of biodiversity across a wide variety of communities. I discuss robustness further in the concluding chapter 10 because I believe that understanding why the neutral theory does so well in spite of manifest differences among species also provides the key to reconciling neutral, dispersal-assembly theory with niche-assembly theory.

SUMMARY

1. There are two important statistical distributions in the unified neutral theory. One is the asymptotic logseries distribution for the metacommunity under point mutation speciation; the other is a new statistical distribution, the zero-sum multinomial, which arises in the local community under dispersal limitation from the metacommunity. The zero-sum multinomial is also the metacommunity relative abundance distribution under random fission speciation.

2. Fitting the fundamental biodiversity number and the dispersal rate currently requires simulations for any reasonable-sized community. Publication of an extensive tabling of the zero-sum multinomial distribution is planned to aid in hypothesis testing.

3. A fast algorithm for computing the expected metacommunity relative abundance distribution under point mutation speciation is given.

4. In general, the fundamental biodiversity number will tend to be underestimated from local samples of

relative species abundance. The underestimation is hard to estimate a priori because it depends on the amount of dispersal limitation. However, better estimates of θ are obtained if the fitting is restricted to the common species in the local community, which are more likely to be widespread metacommunity species. The best estimates of θ are likely to be obtained from pooling samples collected all across the metacommunity.

5. The theory fits the relative species abundance data from a diverse array of communities, from marine planktonic copepods to tropical trees. However, there are many distributions that it does not fit well for expected biological reasons, particularly for violations of the zero-sum rule.

6. Therefore, the theory is expected to be useful in testing the importance of ecological drift, random dispersal, and random speciation in structuring natural communities.

CHAPTER TEN

Reconciling Dispersal-Assembly and Niche-Assembly Theories

The principal aim of this book has been to show that a powerful, formal neutral theory of biodiversity and biogeography can be constructed on the foundation of the original theory of island biogeography. This theory describes the dynamics of species richness and relative species abundance in local communities and metacommunities that are undergoing ecological drift, random dispersal, and random speciation. The theory's key assumptions are, first, that the dynamics of ecological communities are a stochastic, zero-sum game, and second, that this game is neutral and played by individuals that are identical in their probabilities of giving birth, dying, and dispersing—and even of speciating. In chapter 1, I explored the philosophical divide between the dispersal assembly and niche assembly perspectives of ecological communities. In this concluding chapter, I anticipate a truly unified theory that at a more fundamental level reconciles these two apparently conflicting perspectives.

My prediction is that this reconciliation will resemble the unification of population genetics, which successfully combined the theories of genetic drift and of selection. Such a reconciliation in ecology will depend on wider acceptance of the fact that ecological drift is a universally present process that must inevitably operate continuously in real ecological communities, just as genetic drift is always operating on gene frequencies in populations. The appropriate question therefore is not whether ecological drift exists, but under what circumstances is it quantitatively important. I would argue that we really have no idea because we have not looked adequately nor in the right ways. Our half-century preoccupation in ecology with equilibrium theories of coexistence in niche-assembled communities (chapter 1), has led us away from engaging in a thorough examination of the importance of ecological drift, both empirically and theoretically.

I believe that community ecology will have to rethink completely the classical niche-assembly paradigm from first principles. Such a rethinking is important because there are an embarrassingly large number of empirically well-documented and widespread patterns in community ecology that remain almost totally unexplained by contemporary theoretical ecology. These include regularities in the patterns of relative species abundance, species-area relationships, and even phylogeny. The neutral theory outlined in this book does a better job at explaining these patterns than current niche assembly theory does. In fact, many nonobvious patterns predicted by the neutral theory have been substantiated empirically, so the neutral theory is unexpectedly rich in new, testable predictions. This is not to argue that all of these patterns are due to neutrality, but it does shift the burden of proof to those who would argue that niche assembly is necessary to explain these patterns. I firmly believe that a new synthetic theory of biodiversity and biogeography will be developed which integrates ecological drift and demographic and environmental stochasticity into niche assembly theory, but this will require a deep reevaluation of the current niche-assembly theoretical paradigm. There is abundant evidence that such a fundamental rethinking has already begun. A number of ecologists have expressed ideas that are similar but not identical to those presented here (e.g., Levins 1968, Tilman 1988, Charnov 1993).

One of the great surprises to me in developing this theory has been just how well, in fact, it *does* work. It works astonishingly well in spite of making what might appear to be a false and crippling assumption, namely, that all individuals are identical. As I will argue below, however, this assumption is far closer to the truth than might at first be appreciated. Why does the neutral theory do so well? In this chapter, I present my thoughts on the answer to this question, and its implications for the reconciliation of niche- assembly and dispersal-assembly perspectives.

The success of this neutral theory may be distressing to some people, but it may be of some comfort to note that the neutrality assumptions in this neutral theory are far more biologically interesting than simply asserting that "nothing is going on." Indeed, there are many possible neutral models that could be studied that would allow far more complexity in the ecological interactions among individuals than those studied in this book (chapter 1). The essential point to bear in mind is that assuming ecological equivalence among individuals is a very permissive definition of neutrality that potentially allows a great deal of interesting biology to be incorporated in neutral theories. In what follows I discuss the reasons why I think assuming ecological equivalence among individuals in trophically defined communities may not be such an unreasonable approximation of nature after all.

I believe that the success of the neutral theory is powerful testimony to the fact that, in some regards, ecological nature is far simpler than we might at first suppose. If one always had to specify the unique dynamical behavior of each and every species and all their interactions with resources and each other in a community, then the neutral theory simply would not work at all. The very fact that the neutral theory does so well implies that the dynamics of ecological communities are governed by rules that have much lower dimensionality than their species richness might suggest.

The key to reconciling niche assembly and dispersal assembly, I believe, lies in recognizing an initially counterintuitive truth. It is not immediately obvious why, but niche differentiation actually *greatly reduces* the dimensionality of ecological communities. In principle, there is no necessary reason to expect this. However, the niche differentiation of virtually all species in trophically well defined communities is of a special type, confined to a very restricted region of possible life history space. Virtually all niche differentiation is constrained to obey a small set of scaling laws and fitness invariance rules that embody a series of life history trade-offs shared by all species in the community. These trade-offs and fitness invariance rules can be fully specified by many fewer parameters than would be required to uniquely characterize the niche of each and every species.

By *fitness invariance* I mean that there are different trade-off combinations of life-history traits that confer equivalent per capita relative fitnesses on the species exhibiting them. This must be true by an almost self-evident proof. All species that manage to persist in a community for long periods with other species must exhibit net long-term population growth rates of nearly zero. A given species may win a little here, and lose a little there, but over the long term and over large landscapes, its net growth rate must be very close to zero. If this were not the case, i.e., if some species should manage to achieve a positive growth rate for a considerable length of time, then from our first principle of the biotic saturation of landscapes, it must eventually drive other species from the community. But if all species have the same net population growth rate of zero on local to regional scales, then ipso facto they must have identical or nearly identical per capita relative fitnesses.

Consider the tropical forest on Barro Colorado Island (BCI), Panama. In the 50 ha plot there are more than three hundred coexisting tree and shrub species with stems greater than 1 cm in diameter at breast height (dbh). Although the BCI forest is species rich, closer examination shows that virtually all BCI tree species are niche differentiated along a few major life history trade-offs. The best known of these is the trade-off between shade tolerance and growth rate. Shade-tolerant trees have high tolerance (high survival) to shade stress but are unable to grow rapidly in full sun. Conversely, shade-intolerant pioneers are able to grow very rapidly in the high light environment of a large treefall gap, but they have very poor tolerance of shade (low survival). This life history trade-off can be visualized graphically by plotting mean annual survival rate of a tree species in the shade against median annual growth rate in full sun. The trade-off is manifest as a narrow locus of points in niche space along which all BCI tree species are arrayed (fig. 10.1). Most of the tree species in the BCI forest are shade tolerant, and so there are more species at the shade-adapted end of the manifold than at the sun-adapted end. The physiological basis for this trade-off is well known (e.g., Bazzaz and Pickett 1980, Coley et al. 1985, Reich et al. 1999). This is an example of what has been called a *life history manifold* because it is an evolutionary attractor in niche space. It represents a series of unavoidable or nearly unavoidable life history design or constraint functions that every species in the trophically defined community must obey in order to stay in the game. Clearly, this manifold greatly reduces the dimensionality problem of characterizing life-history variation in the BCI tree community. The manifold can be described by many fewer parameters than could each and every tree species individually. With two parameters, one can characterize the principal linear axis, and with a few more parameters, one can describe even the distribution of species densities along the manifold.

Moreover, niche differentiation in the BCI tree community is further simplified because a majority of the other major life history traits of BCI tree species also covary predictably along the shade tolerance-intolerance manifold. For example, shade-tolerant species in general produce fewer, larger seeds that disperse relatively locally and germinate quickly in understory shade. Shade-intolerant pioneer species tend to produce large numbers of small seeds that

The irony of this argument in the context of the present neutral theory is that it turns out that *niche differentiation along life history trade-offs is the very mechanism by which per capita relative fitnesses are equalized among the coexisting species in a community.* In light of this, the question that we started with, namely, "Which is more important, niche assembly or dispersal assembly?" now appears somewhat simplistic because these two theoretical perspectives are fundamentally and intrinsically intertwined and interdependent mechanistically. This logic leads to a set of perhaps deeper and subtler questions.

If per capita relative fitnesses are equalized by life-history invariance rules, then why don't "cheater" species invade the communities that break these rules by escaping the life-history trade-offs? The answer is that there probably *are* cheaters from time to time that manage to break the rules. Most of the time, however, the microevolutionary change we observe in non-rule-breaking species is constrained to take place along preexisting life-history trade-offs. Thus, the mean phenotype of a species may slide up or down along the current community life-history manifold, all the while maintaining fitness invariance. The current position of the species on the life-history manifold will reflect its history of selection in past environments. For example, a tree species might gradually evolve higher growth rates as a result of a history of exposure to higher light environments, but in so doing it would be expected to lose the correlated traits for shade tolerance. Fitness invariance arising from adaptive trade-offs is not a new argument, which was perhaps earliest and best expressed by Richard Levins (1968) in his now classic book, *Evolution in Changing Environments.* Among other things, Levins argued that adaptive trade-offs are inevitable, and escaping them is very difficult, in part because organisms are limited by the amount of genetic information they can carry (preventing the evolution of superorganisms that are best in all environments), and in part because of logical inconsistencies (e.g., being simultaneously large and small, many seeded and few seeded), historical-morphogenetical constraints, and ultimately, boundary conditions on the possible imposed by physico-chemical laws.

Every now and then, however, a species does manage to break partially free of the constraints of the life-history manifold currently governing its community, and this species will achieve a new level of fitness with a somewhat different set of constraints. For a time, such a species will be a superior competitor, sweeping communities free of, or at least reducing, populations of its competitors, depending on the magnitude of differences in relative fitness. Introduced exotic species that are successful invaders of biotically saturated communities are often such rule breakers in their new community settings. However, this state of affairs cannot go on for long, ecologically or evolutionarily. The existence of rule breakers establishes a new, higher fitness criterion for all other species in the community, which then come under strong selection for life-history adjustments that increase their fitness to match the rule breaker. Presumably, minor evolutionary rule breaking goes on all the time but is largely undetected. However, large breaks in the rules that are also successful are presumably increasingly infrequent the larger the break in the rules is. This is analogous to the argument behind David Raup's (1991) "kill curve" for the distribution of sizes of extinction events, but in reverse. Thus, it is only when a massive rule-breaking episode takes place in the fossil record that we identify and label it as an "adaptive radiation."

In summary, although there is no necessary reason why niche differentiation in principle should simplify community dynamics, it usually does. But why? After all, niche differentiation makes species *different*, doesn't it? I believe the answer to this question is both *yes* and *no*. The answer is *yes* in the sense that species do exhibit different life-history traits that allow, among other things, taxonomists and ecologists to distinguish one species from another. But the answer is *no* in the sense that these niche differences obey life-history trade-off rules that maintain per capita fitness equivalence among the niche- differentiated species.

Niche differences follow life-history invariance rules that maintain equal relative fitnesses, but this only takes us halfway to reconciling the niche assembly with dispersal-assembly perspectives. Now we must ask: What determines the species richness and relative species abundance of the species in the community? According to the neutral theory, mostly what mechanisms remain once per capita relative fitnesses are equalized are processes of ecological drift. This implies that *life-history trade-offs and fitness invariance rules potentially decouple niche differentiation from control of the species richness and relative species abundance of communities.*

Once species' fitnesses are equalized, the degree of decoupling of niche differentiation from relative abundance will depend on the degree to which zero-sum dynamics apply. I argued in chapter 3 that the zero-sum rule must govern the dynamics of many communities of trophically similar species. I have given many examples in this book of patterns of relative species abundance that are qualitatively and quantitatively consistent with zero-sum ecological drift in a diverse array of communities. But the question remains: How generally do zero-sum dynamics apply among species on the same trophic level? The neutral theory makes the simplifying assumption that any species in a community of trophically similar species has the potential to become mon-odominant, occupying all available space or using all limiting resources, if no other competing species are present. In other words, no limiting resource will go unused for long if we vary the species richness and composition of a community. Thus, the zero-sum rule assumes that resources limiting the community are available to all member species of the community. This assumption will not always be true, or it may be true only on evolutionary timescales. However, once again I argue that we really have no idea because we have not been asking the right questions. The phenomenon of "ecological release" noted on islands (e.g., Crowell 1962, 1973) may actually be far more prevalent in natural communities than we acknowledge. Indeed, it may be going on all around us all the time on a micro scale.

I believe that species' adaptability and malleability in resource use are why the assumption of zero-sum dynamics works so well in so many trophically defined communities. In many cases, any species in the community can use any of the resources that limit the total community. Resources normally used by one species are often opportunistically used by another species whenever the former species happens to be absent. Species may use some resources slightly more or less efficiently, but these differences are unlikely to seriously violate the zero-sum rule.

Opportunism probably happens far more often than current niche theory would lead us to believe. There is a pervasive tendency in ecology to treat all individuals of a given species as exhibiting the mean species phenotype, i.e., to treat species typologically as invariant entities. We routinely overlook the flexibility, opportunism, and facultative use of resources of which variable individuals in species are universally capable—not to mention potential evolutionary responses. I believe that most species will be found to be far less specialized in resource use than current theory suggests. Virtually all species have at least some capability of switching resources if their preferred resources are unexpectedly scarce. I think that the rigidity of our current niche typology has blinded us to not seeing the pervasive facultative and opportunistic use of resources by species. I therefore predict that zero-sum dynamics will apply far more generally among species on a given trophic level than one would predict based on current theory. This rigid niche typology is also what has led us to the prediction that competitive exclusion will be commonplace when in fact it is rarely if ever observed in nature.

Thinking typologically about species also dramatically affects our ideas about speciation and the meaning of biodiversity itself. We have seen that the unified theory generally fits patterns of biodiversity best under the assumption that species are fractal entities. According to this view, species and higher taxa are vessels of self-similar phenotypic and genotypic variation on biological scales running from the individual level all the way up to the metacommunity level of organization (chapter 8). The "point mutation" mode of speciation is more consistent with the data and with a fractal view of species, and this is the theory that generates the fundamental biodiversity number θ. The "random fission" mode of speciation arises from the classical model of allopatric speciation (Mayr 1963), but at this preliminary point in time, this mode appears to be less consistent with the data (chapter 8). If we accept a fractal view of species and of the variation that they encompass, then understanding the true meaning of biodiversity is a different and more challenging problem than we might have thought.

In order to test the unified neutral theory, therefore, we need much more research on the extent to which species in trophically well-defined communities obey the zero-sum rule, as well as research on the mechanisms of variable resource use by variable individuals within and among variable species. Such questions were once pursued vigorously more than two decades ago in studies of within- and between-phenotype niche breadths (e.g., Roughgarden 1972, Colwell and Futuyma 1971), but these are still fundamental, unresolved questions on which we need to reinvigorate research.

One of the ways that the zero-sum rule breaks down is in the case of habitat specialization. In developing the present neutral theory, I assumed that there was a completely homogeneous habitat across the metacommunity landscape, and I asked what patterns of biodiversity and biogeography would develop under ecological drift, random dispersal, and random speciation. However, in nature real landscapes consist of environmental gradients, and very different habitats exist in a landscape mosaic. One can develop alternative theories for the maintenance of biodiversity on heterogeneous landscapes (Tilman 1988, Tilman and Pacala 1993, Ritchie and Olff 1999). Moreover, the existence of environmental gradients that affect relative fitness of species also lead to important complexities such as source and sink populations (Pulliam 1988). All of these ideas are worthy of much more theoretical development, particularly in conjunction with the theory of ecological drift, but these developments lie in the future, and are beyond the scope of the present work.

I would like to conclude this book with a discussion and response to a number of critiques that have appeared of early versions of the theory. I will respond in detail to only two of these—one empirical, and one theoretical—because my responses make some general points that are germane to the question of how to reconcile niche assembly and dispersal assembly theories.

Terborgh et al. (1996) analyzed several plots of mature tropical forest, each 1–2.5 ha, along a 40 km stretch of the Manu River in Amazonian Peru. They argued that the abundance rankings of the top twenty commonest species were too similar from plot to plot and over such a distance to be consistent with ecological drift (which they called the "nonequilibrium theory"). Pandolfi (1996) (also see Jackson et al. 1996 and Pandolfi and Minchin 1995) analyzed a continuous chronosequence of fossil reef terraces gradually uplifted along the north coast of Papua New Guinea. The record extends over a period of 95,000 years through repeated sea level and surface temperature changes, and also extends spatially for tens of kilometers along the coast. Pandolfi found relative constancy in both taxonomic composition and in species richness, punctuated by intervals of relatively rapid change, over this period of nearly 100,000 years. They attributed the long periods of relative stability to the existence of limited-membership reef communities that were noninvasible (i.e., niche assembled).

The conclusions of Terborgh et al. (1996) and Pandolfi (1996) are premature, according to the unified theory, for a number of reasons. First, we know from the theory presented in previous chapters that common and widespread metacommunity species are likely to be very resistant to extinction and to persist for geologically significant periods of time. Second, in chapters 4 and 7 we showed that very modest rates of dispersal will insure that these common metacommunity species are nearly everywhere all the time. Third, increasing the rate of dispersal will increase the proportion of metacommunity diversity that is present in local communities.

Attempting to test ecological drift theory using only common metacommunity species is particularly problematic. For one thing, there is a positive correlation between the total abundance of species and their geographic ranges (Brown 1984, 1995). Limiting one's attention to the most abundant species is to bias the analysis in favor of those species that are least likely to be dispersal limited and most likely to be persistent in space and time. Therefore, one's conclusions are strongly biased toward constancy of community composition. We have seen that the time to extinction $T(N_i)$ of a very common metacommunity species is expected to be extremely long (chapters 4, 8). From θ we know that J_M is an enormous number–on the order of the inverse of the speciation rate v. But the time to extinction $T(N_i)$ for large N_i is an even bigger number, on the order of the metacommunity size times the log of metacommunity size, times the initial population size of the species (chapter 4). The immensity of this number is important because it means that common species will be very persistent members of the metacommunity and of local communities as well. This number is so large that it means that the ecological dynamics of metacommunities are temporally commensurate with

331

the evolutionary dynamics of speciation and extinction. Indeed, this has to be so because we have proved that a steady-state metacommunity biodiversity is established at equilibrium between speciation and extinction. Paleobiological evidence that widespread species are more resistant to extinction events has been provided by Jablonski (1995) and Jackson (1995).

These long persistence times mean that very abundant metacommunity species have plenty of time to disperse nearly everywhere. Community stability should therefore be increasingly conspicuous as one looks at aggregate species abundances at higher and higher taxonomic levels. This effect was noted by Gentry (1982) in a large number of 0.1 ha forest plots inventoried all over tropical South America. Gentry found that the familial composition of tropical forests is extremely constant, which is exactly what the unified theory would predict. This stability is shown in figure 10.2, which presents the family level composition of the BCI forest, for all of Panama, and for the world. The cumulative distributions of families are nearly identical from local to global scales. Results such as these can also be taken as support for the niche assembly perspective. If the early radiation of the angiosperms produced a rapid filling of niches at the family level, and if these niches are ubiquitous in plant communities, then this result would be expected. Thus, the problem is that both theories of community organization—dispersal assembly and niche assembly—predict the same results, and they certainly do not falsify ecological drift.

In chapter 7 I discussed incidence functions, which give the probability that a species will be present in a local community or habitat patch as a function of patch size J and the probability of immigration per birth m. Recall the relatively conservative case of a species that is only moderately common in the metacommunity, constituting just one percent of it (fig. 7.2). In spite of relative rarity in the metacommunity

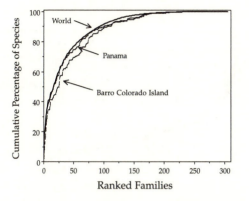

FIG. 10.2. High degree of constancy in the family-level composition of floras from local scales (BCI) to global spatial scales. Such a result is predicted by both the unified theory and by niche assembly theory, and therefore is not useful in testing them.

as a whole, this species will still be essentially always present (100%) in a local community of 10^3 individuals or more if $m \geq 0.1$, in communities of 10^4 individuals or more if $m \geq 0.01$, and in communities of 10^5 individuals or more if $m \geq 0.001$. In this context, let us reconsider the Terborgh example. Suppose that in round numbers there are 10^4 trees >10 cm in trunk diameter at breast height in a 25 ha plot of Manu forest (estimate based on BCI). Now let us conservatively imagine a corridor of forest 40 km long consisting of 1 km² plots along the Manu River, each containing approximately $4 \cdot 10^4$ trees. Therefore, the corridor contains about $1.6 \cdot 10^6$ trees. On BCI we estimated the immigration rate for the 0.5 km² plot to be close to 0.1. Suppose that for a square kilometer the value of m is half that value, or 0.05. For the moment, assume that species must migrate only within this corridor. This is a more restrictive case than reality because the forest extends in all dimensions for much greater distances. Let us now simulate the Manu metacommunity and measure the predicted presence of the twenty

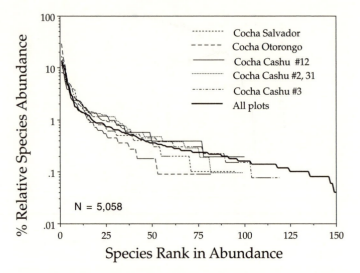

FIG. 10.3. Dominance-diversity curves for five plots of 1 to 2.5 ha in mature forest along the Manu River in Amazonian Peru. Three of the plots are nearly contiguous with one another (the plots at Cocha Cashu) and separated by about 40 km from the other two sites (Cocha Salvador and Cocha Otorongo), which were close together. The "all plots" is a composite of all five plots. These data represent all species with a total abundance of four individuals or more. Unpublished data of John Terborgh.

most abundant species in local community samples of one thousand trees.

Using unpublished raw data on Manu tree abundances provided by John Terborgh, I plotted the dominance-diversity curves for the five plots discussed in Terborgh et al. (1996) (fig. 10.3). Three of the plots were nearly contiguous at a site named Cocha Cashu and were approximately 40 km from the other two sites (Cocha Salvador and Cocha Otorongo), which were near to each other. Because of potential problems with taxonomy, I excluded from the analysis any species with fewer than four individuals in a plot, with one exception. One of the plots was less than half the size of the others (trail 12), and in this case I included all species with two or more individuals (to attempt to equal-ize sampling intensity). In chapter 6, I demonstrated that pooling a series of small samples taken across the metacom-munity should help compensate for dispersal limitation and give a better estimate of the metacommunity relative abun-dance distribution. Therefore, I pooled all five plots for an estimate of the species richness and relative species abun-dance distribution of the Manu metacommunity. Four of the five sites have steeper dominance-diversity curves than the metacommunity aggregate (fig. 10.3), as predicted by dis-persal limitation in the unified theory, and of course fewer species.

The simulations show that the top sixteen species are expected to be present in all local stands essentially 100% of the time (fig. 10.4). The Manu forest is heavily domi-nated by two palm species, *Iriartea deltoides* and *Astrocaryum macrocalyx*, which together make up a quarter (24.5%) of all trees in the forest. The top twenty species in abundance col-lectively make up two-thirds (65.1%) of the forest. Also as expected from the theory, the average rank abundance of the species was the same in the local stands as in the meta-community, although the actual rank positions in individual stands varied. The variation in rank position among local stands increased for the higher-ranked, rarer species, which is exactly what Terborgh et al. (1996) observed.

Ironically, Terborgh et al. (1996) may have been correct about the importance of niche assembly in the Manu forests for at least one reason that they did not give. They were correct in noticing that the dominants were everywhere and nearly equally dominant in rank position in all plots. But this observation alone does not contradict ecological drift, as we have just pointed out. What they did not comment on, however, was the excessive dominance itself—presumably because they had no prior statistical hypothesis of what null relative abundance distribution to expect. The unified the-ory detects significant levels of dominance in the seven most

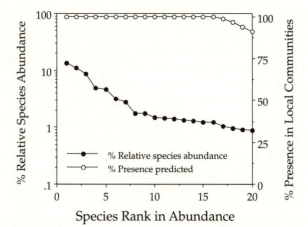

FIG. 10.4. The percentage relative species abundance of the twenty most abundant species in five mature forest plots along the Manu River in Amazonian Peru, and the percentage of plots of comparable size in which presence of the ranked species is predicted. The closed circles are the observed percentage relative species abundances. The open circles are the predicted percentages of plots that will contain the species. Unpublished data provided by John Terborgh.

abundant species over that predicted by the best-fit meta-community relative abundance distribution. Note that there is a kink and sudden upward bend in the distribution at a rank abundance of eight (fig. 10.4). This implies that the top seven species enjoy some relative fitness advantage over the remaining species in the metacommunity. This advantage might be expressed as a reduced per capita probability of death or a higher per capita probability of birth, or in some combination of the two.

The exact biological reasons for the excessive dominance in the top seven most abundant species in the Manu forest are currently unknown, so it is not presently possible to model them. However, I can illustrate how easily in principle one can modify the stochastic equations of zero-sum ecological drift to incorporate a fitness advantage for given species. Suppose, for example, we assign each of the top seven species a slight competitive advantage by increasing its per capita survival rate over other species. Under pure drift, the probability of a death in the ith species is N_i/J. Now let s_i be the survival advantage of the ith species. Then the probability that the next death in the community will occur to species i is simply $N_i(1 - s_i)/J$. Note the obvious parallels with the way in which selection is incorporated in population genetics models. One can fit the s_i by a combination of simulation and maximum likelihood procedures.

I have done this for the Manu forest example, and the results are presented in figure 10.5. Several conclusions from this case study are interesting and of general significance. The most important general result is that *incorporating modest fitness differences does not result in the rapid competitive exclusion and extinction of disadvantaged species in the metacommunity*. This result is not predicted by classical theory, which instead predicts that competitive exclusion will take place quickly, even with relatively slight competitive differences (e.g., Zhang and Lin 1997). This difference from classical theory is due to dispersal limitation, which almost indefinitely delays competitive exclusion (Tilman 1994, Hurtt and Pacala 1995). I will explain this result later. It has important general implications for community ecology, as well as for the ultimate reconciliation of dispersal assembly and niche assembly theory.

The second conclusion, which is also general, is that *relatively small fitness differences can nevertheless produce large differences in steady-state relative species abundance*. In the Manu case, for example, I calculate that the commonest species has a fitness advantage of only about 6%, but this small advantage resulted in almost a doubling of the relative abundance of the rank-1 species over what was expected under pure ecolgical drift (fig. 10.5). The remaining six species with elevated abundances have even smaller relative fitness advantages. In general, this and other case studies indicate that the unified theory is capable of detecting small differences in fitness.

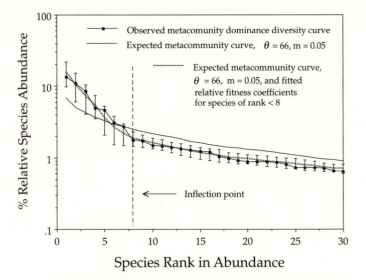

FIG. 10.5. Dominance-diversity curve for trees in old-growth forest along the Manu River in Manu National Park, Peru. The thin line without symbols is the expected metacommunity distribution under pure ecological drift. The dotted line is the expected metacommunity distribution with survival advantages for the seven most abundant species. The observed dominance diversity is the curve with filled-circle symbols. Only very slight survival advantages were required to fit the curve. The largest advantage was just 6% for the rank-1 species, which resulted in a nearly twofold increase in steady state abundance over the abundance expected under pure zero-sum ecological drift. The error bars are ±1 standard deviation of the observed mean.

I suggest using the term *ecological dominance deviation* for cases in which dominance in ecological communities statistically exceeds that predicted under ecological drift.

The third conclusion is an empirical generalization that so far has held in most communities I have examined, and is true in the Manu case, but its acceptance awaits much further testing. I tentatively conclude that *ecological dominance deviations, if present at all, are usually detectable in only a small fraction of the species in a community.* In the Manu forest case, for example, the relative abundance distribution of the species remaining after the most abundant seven species are taken out is statistically indistinguishable from that expected under pure community drift. It is as if the remainder of the metacommunity distribution is collectively downshifted to lower relative abundances. Note that the observed curve is almost perfectly parallel to the expected metacommunity curve under pure drift for the species of rank-8 and higher (fig. 10.4). This suggests that ecologists should explore mixed models of ecological communities in which a few species are competitive dominants that sequester most of the resources under the zero-sum game, and that the remaining species in the community are gleaners that compete for the leftovers under a more neutral zero-sum game. These ideas are reminiscent of the niche preemption theories of Motomura (1932) and Whittaker (1965). Once again, the competitive dominants cannot take over completely, according to the unified theory, because of steady-state dispersal limitation.

The final conclusion is really a corollary of the second and third conclusions, but it bears emphasizing. Even when ecological dominance deviations are detectable, the assumption of per capita equivalence among all species is rarely if ever quantitatively far from the truth. Thus, even when some species in the community have detectable competitive advantages, nevertheless these advantages are often quite small and therefore often do not seriously compromise the assumptions of neutrality or the zero-sum game.

Before leaving the subject of detecting ecological dominance deviations, I offer one theoretical example. Figure 10.6 shows the effect of giving the rank-1 species a 25% and a 50% survival advantage, respectively, over all other species in the community. I examined a case of a local community of size $J = 1600$, which receives immigrants with probability $m = 0.1$ per birth from a metacommunity with a fundamental biodiversity number $\theta = 25$. I assume that the dominant species is locally dominant only, so that

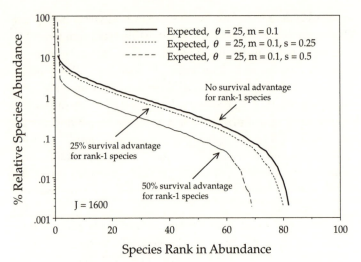

Fig. 10.6. The effect of ecological dominance on the theoretical distribution of relative species abundance in a local community of size $J = 1600$ under moderate dispersal limitation ($m = 0.1$). Ecological dominance in this case is due to a survival advantage in the rank-1 species of either 25% or 50%. Note that the curves for the higher-rank species remain parallel to the expected curve under pure drift. This is expected if all remaining species engage in a neutral zero-sum game.

its relative abundance in the metacommunity is as expected under pure drift. With no advantage, the expected local abundance of the rank-1 species with moderate dispersal limitation ($m = 0.1$) is a little over 10%. This percentage relative species abundance rises to 28% for a survival advantage of 25%, and to 71% for an advantage of 50% (fig. 10.6).

There are other kinds of significant ecological dominance deviations that might also be observed, not just in the most abundant species, but also in species of middle abundance rank or rare species. For example, in the previous chapter we illustrated two cases of such deviations, one for the bird community in 10 ha of tropical forest in Manu National Park (Terborgh et al. 1990; fig. 9.13), and the other for a rain-forest-canopy beetle community (Stork 1997; fig. 9.17). In the case of the Manu bird community, there was a hump in the dominance-diversity curve at intermediate abundances, as well as excess dominance in the most abundant species. In the case of the canopy beetle community, there were too many species represented by singletons. In each case, we were able to detect significant departures from the neutral expectation.

I turn now to a theoretical paper that is critical of the theory of ecological drift. I will argue that understanding and answering the authors' objections goes a long way toward solving the larger problem of reconciling the dispersal assembly and niche assembly perspectives on the organization of ecological communities.

Zhang and Lin (1997) argued that the long coexistence times that are expected for identical species under zero-sum ecological drift (called "community drift" in Hubbell and Foster 1986b) is a fragile result. They introduced small differences in fecundity into the absorbing case of the ecological drift model, and found that relatively small differences in fecundity led to a dramatic shortening of the expected mean lifespans of species. They studied a two-species community in which the species were identical in all respects except for per capita fecundity. Let w be the per capita fecundity factor of the focal species relative to the nonfocal species. Thus, if $w = 1$, then the per capita fecundities of the two species are identical.

Now consider the modified transition probabilities for the focal species that take into account the differential per capita fecundity; note that the equations reduce to the absorbing case discussed in chapter 4 if $w = 1$.

$$\Pr\{N_i - 1 | N_i\} = \left(\frac{N_i}{J}\right)\left(\frac{J - N_i}{(N_i - 1)w + J - 1}\right)$$

$$\Pr\{N_i | N_i\} = 1 - \Pr\{N_i - 1 | N_i\} - \Pr\{N_i + 1 | N_i\}$$

$$\Pr\{N_i + 1 | N_i\} = \left(\frac{J - N_i}{J}\right)\left(\frac{N_i w}{J - 1 + N_i(w - 1)}\right).$$

To simplify notation, let $p_n = \Pr\{N_i + 1|N_i\}$ and $q_n = \Pr\{N_i - 1|N_i\}$. The mean time to extinction or complete dominance for the focal species, starting at abundance N_i, can be found analytically (Zhang and Lin 1997):

$$T(N_i) = \frac{\sum_{k=0}^{J-1} R_k}{\sum_{k=0}^{J-1} e_k} \sum_{k=0}^{N_i-1} e_k - \sum_{k=0}^{N_k-1} R_k \quad \text{for } 0 < N_i < J,$$

where

$$e_k = \frac{q_1 q_2 \cdots q_k}{p_1 p_2 \cdots p_k}, e_0 = 1,$$

$$R_k = \frac{1}{p_k}\left[1 + \frac{q_k}{p_{k-1}} + \ldots + \frac{q_k \cdots q_2}{p_{k-1 \ldots p_1}}\right], \text{ and } R_0 = 0.$$

Zhang and Lin (1997) then studied how the time to extinction or complete dominance in a two-species community was affected by the differential fecundity factor w. There were rapid decreases in the time to "fixation" as w increased to about 1.2. However, for larger values of w, the rate of decline in species lifetimes was much slower (fig. 10.7). Zhang and Lin focued their attention on the fast decline in the time to fixation for small differences in relative fecundity, but they failed to note that the species lifetimes remained very large. More important, there are still huge increases in the time to fixation as community size J increases. They only computed lifetimes in small communities below 4096 individuals. As community size grows as large as a naturally occurring metacommunity, average species lifespans will become very long, even under the condition of unequal fecundities. This is one of the main conclusions of chapters 5 and 8, which show that the times to extinction will be so long that a diversity equilibrium will be established with the speciation rate.

Perhaps the most substantive criticism of Zhang and Lin's (1997) result is that their model fails to take dispersal and dispersal limitation into account. In chapter 4 we also considered a model of local community dynamics in the absence of dispersal limitation. In that model, as in Zhang and Lin's,

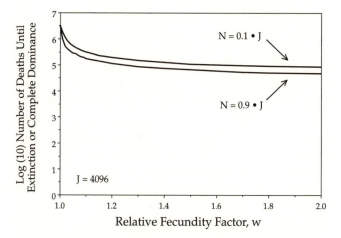

FIG. 10.7. Effect of competitive asymmetry (unequal fecundity) on the expected time to extinction or monodominance in the model community of Zhang and Lin (1997). Note the rapid decrease in mean lifespan with increasing difference in relative fecundity, especially for $w < 1.2$. Note also, however, that the rate of decline in lifespan slows down considerably for higher w. Two cases are drawn, one for a species having an initial abundance of 10% of the community, and one for a species having an initial abundance of 90% of the community. Redrawn from Zhang and Lin (1997).

each individual was equally likely to give birth to the replacement for any dying individual, regardless of whether the death occurred nearby in the same local community or far away in some other community in the metacommunity. However, chapters 6 and 7 of this book have largely been devoted to exploring the consequences of dispersal limitation under ecological drift for equilibrium species-area relationships and metacommunity biodiversity. According to the unified theory, these large-scale spatial patterns of biodiversity arise through an interaction of slow speciation rates, long extinction times of abundant and widespread metacommunity species, and steady-state limitation in the rate of dispersal of species over the metacommunity landscape.

The significance of dispersal limitation to the coexistence of species in ecological communities has been studied explicitly in two important theoretical papers by Tilman (1994) and Hurtt and Pacala (1995). Tilman (1994) showed that if there was a strict, ordered trade-off between competitive ability and dispersal ability, such that the best competitor was the worst disperser and vice versa, then in principle there is no upper limit to the number of coexisting species. Hurtt and Pacala (1996) went further and demonstrated that dispersal limitation all by itself can overwhelm the effects of even strong competitive differences, especially in species-rich communities. They analyzed a model of a community of niche-differentiated species partitioning an environmental axis. Each species was the absolutely best competitor in some region of the axis. Without dispersal limitation, each species as expected won those sites in which it was the best competitor. With dispersal limitation, however, many sites were won by "forfeit" by species that were not the absolutely best competitor for the site.

Hurtt and Pacala (1995) also demonstrated that a runaway process can occur between species richness and community-wide dispersal limitation. As communities become richer in species, the average population density of individual species falls, increasing the mean steady-state level of dispersal limitation in the community. This increases the number of forfeited sites and the number of species that win by default. The result is that the exclusion of inferior competitors becomes nearly indefinitely delayed. This runaway process is more likely to start if competitive differences among species are not large, but once it has begun, the effects of dispersal limitation can overwhelm even large competitive differences. The limits to community species richness and relative species abundance are then set, not by local niche differentiation, but by regional biogeographic processes—speciation, extinction, migration, and drift on the metacommunity landscape.

The true significance of these two papers is that they provide yet another mechanism that largely decouples niche differentiation from the control of species richness and relative species abundance in ecological communities. Among other things, these papers imply that a complete continuum is possible in the quantitative degree to which niche assembly versus dispersal assembly controls local community structure and relative species abundance.

In this book I hope that I have achieved my three original goals, which were, first, to explore how far we can get with a formal neutral theory in community ecology; second, to show that both ecological drift and biotic interactions are important to the assembly and dynamics of ecological communities; and, finally, to reduce the pervasive resistance of ecologists to neutral theory by demonstrating its considerable power and the richness of its testable predictions. Much theoretical and empirical work remains to be done, however, to achieve a truly synthetic theory of biodiversity and biogeography, and I look forward with considerable excitement to continued progress on both fronts.

At the end of their monograph on the theory of island biogeography, MacArthur and Wilson (1967) offered a brief prospect for the future of the science of biogeography. "Biogeography appears to us," they wrote, "to have developed to the extent that it can be reformulated in terms of the principles of population ecology and genetics .... To achieve this restatement ... requires both theory and experiments that must be in large part novel. Simultaneously it demands a cultivation of population and community ecology in a way that contains much more evolutionary interpretation than has been traditional."

It is my hope that the unified neutral theory of biodiversity and biogeography that has been presented here—a theory that incorporates speciation and relative species abundance into the original theory of island biogeography—represents

345

a significant step, close to 40 years later, toward achieving MacArthur and Wilson's ambitious goal.

SUMMARY

1. The niche differentiation of virtually all species in natural communities obeys a small set of scaling laws and fitness invariance rules expressed as a series of shared life-history trade-offs.
2. These life-history trade-offs equalize the per capita relative fitness of species in the community, which sets the stage for ecological drift.
3. The equalized per capita relative fitnesses of species potentially decouple niche differentiation from control of species membership and relative abundance in the community.
4. The degree of decoupling and the importance of ecological drift then depend on the degree to which zero-sum dynamics apply to the species in the community.
5. Critics who conclude that natural communities are too stable for the theory, or conversely, that theoretical communities are too unstable under ecological drift, underestimate the importance of dispersal and dispersal limitation in structuring communities.
6. The present unified neutral theory of biodiversity and biogeography is only a beginning. There are many promising new theoretical and empirical paths to explore for achieving the ultimate reconciliation and unification of niche-assembly and dispersal-assembly perspectives in community ecology.

Literature Cited

Andrews, M. J., and D. G. Rickard. 1980. Rehabilitation of the inner Thames estuary. *Marine Pollution Bulletin* **11**: 327–332. (Chap. 9)

Anscombe, F. J. 1950. Sampling theory of the negative binomial and logarithmic series distributions. *Biometrika* **37**: 358–382. (Chap. 9)

Arnold, M. L. 1997. *Natural Hybridization and Evolution.* Oxford University Press, New York. (Chap. 5)

Arrhenius, O. 1921. Species and area. *Journal of Ecology.* **9**: 95–99. (Chap. 6).

Avise, J. C. 1999. *Phylogeography: The History and Formation of Species.* Harvard University Press, Cambridge, MA. (Preface).

Baker, R. R. 1985. Moths: Population estimates, light traps and migration. Pp. 188–211 in L. M. Cook, ed., *Case Studies in Populatin Biology.* Manchester University Press, Manchester, UK. (Chap. 4)

Barigozzi, D. 1982. *Mechanisms of Speciation.* Liss, New York. (Chap. 5)

Bazzaz, F. A., and S.T.A. Pickett. 1980. Physiological ecology of tropical succession: A comparative review. *Annual Review of Ecology and Systematics* **11**: 287–310. (Chap. 10)

Bell, G. 2000. The distribution of abundance in neutral communities. *American Naturalist* **155**: 606–617. (Chap. 3)

Bird, N. M. 1998. *Sustaining the Yield: Improved Timber Harvesting Practices in Belize, 1992–1998.* Natural Resources Institute, University of Greenwich, NRI Catalogue Services, CAB International, Wallingford, UK. (Chap. 5)

Bliss, C. L., and R. A. Fisher. 1953. Fitting the binomial distribution to biological data and a note on the efficient fitting of the negative binomial. *Biometrics* **9**: 176–206. (Chap. 9)

Bonsdorff, E., and V. Koivisto. 1982. The use of the log-normal distribution of individuals among species in monitoring zoobenthos in the northen Baltic Archipelago. *Marine Pollution Bulletien* **13**: 324–327. (Chap. 9)

Boswell, M. T., and G. P. Patil. 1981. Chance mechanisms generating the logarithmic series distribution used in the analysis of number of species and individuals. Pp. 99–130 in G. P. Patil, E. C. Pielou, and W. G. Waters, eds., *Spatial Patterns and Statistical*

347

Distributions. Vol. 1. of *Statistical Ecology.* Pennsylvania State University Press, University Park, PA. (Chap. 2)

Bramson, M., J. T. Cox, and R. Durrett. 1996. Spatial models for species area curves. *Annals of Probability* **24**: 1721–1751. (Chap. 6)

Braun, L. 1950. *Deciduous Forests of Eastern North America.* Blakiston, Philadelphia. (Chap. 9)

Brett, C. E., and G. C. Baird. 1995. Coordinated stasis and evolutionary ecology of Silurian to middle Devonian faunas in the Appalachian Basin. Pp. 285–315 in D. H. Erwin and R. L. Anstey, eds., *New Approaches to Speciation in the Fossil Record.* Columbia University Press, New York. (Chap. 1)

Brown, J. H. 1984. On the relationship between abundence and distribution of species. *American Naturalist* **124**: 225–279. (Chap. 10)

Brown, J. H. 1995. *Macroecology.* University of Chicago Press, Chicago. (Chaps. 1, 3, 9, 10)

Brown, J. H., and A. C. Gibson. 1983. *Biogeography.* Mosby, St. Louis. (Chaps. 1, 6)

Brown, J. H., and A. Kodric-Brown. 1977. Turnover rates in insular biogeography: Effect of immigraton on extinction. *Ecology* **58**: 445–449. (Chap. 1)

Brown, J. H., and M. V. Lomolino. 1989. Independent discovery of the equilibrium theory of island biogeography. *Ecology* **70**: 1954–1957. (Chap. 1)

Brown, W. L., and E. O. Wilson. 1956. Character displacement *Systematic Zoology.* **5**: 49–64. (Chap. 1)

Bulmer, M. G. 1974. On fitting the Poisson lognormal to relative abundance data. *Biometrics* **30**: 101–110. (Chaps. 2, 9)

Burky, T. V. 1995. Extinction rates in archipelagos: Implications for populations in fragmented landscapes. *Conservation Biology* **9**: 527–541. (Chap. 7)

Caswell, H. 1976. Community structure: A neutral model analysis. *Ecological Monographs* **46**: 327–354. (Chaps. 1, 2, 3)

Caswell, H., and J. E. Cohen. 1991. Disturbance and diversity in metapopulations. *Biological Journal of the Linnean Society, London* **42**: 193–218. (Chap. 1)

Charnov, E. L. 1982. *The Theory of Sex Allocation.* Princeton Monographs in Population Biology. Princeton University Press, Princeton, NJ. (Chap. 1)

Charnov, E. L. 1993. *Life History Invariants: Some Explorations of Symmetry in Evolutionary Biology.* Oxford University Press, New York. (Chap. 10)

Chesson, P. L. 1986. Environmental variation and the coexistence of species. Pp. 240–256 in J. Diamond and T. J. Case, eds., *Community Ecology*. Harper and Row, New York. (Chap. 3)

Chesson, P. L., and N. Huntly. 1997. The roles of harsh and fluctuating conditions in the dynamics of ecological communities. *American Naturalist* **150**: 519–553. (Chap. 1)

Chesson, P. L., and R. R. Warner. 1981. Environmental variability promotes coexistence in lottery competitive systems. *American Naturalist* **117**: 923–943. (Chap. 3)

Clements, F. E. 1916. Plant succession: Analysis of the development of vegetation. Carnegie Institution of Washington, publ. no. 242. (Chap. 1)

Cohen, J. E. 1978. *Food Webs and Niche Space*. Monographs in Population Biology. Princeton University Press, Princeton, NJ. (Chap. 1)

Coleman, B. D. 1981. On random placement and species-area relations. *Mathematical Biosciences* **54**: 191–215. (Chap. 1)

Coley, P. D., J. P. Bryant, and F. S. Chapin III. 1985. Resource availability and plant anti-herbivore defense. *Science* **230**: 895–899. (Chap. 10)

Colwell, R. K., and J. A. Coddington. 1995. Estimating terrestrial biodiversity through extrapolation. Pp. 101–118 in D. L. Hawksworth, ed., *Biodiversity, Measurement, and Estimation*. The Royal Society. Chapman and Hall, London. (Chap. 6)

Colwell, R. K., and D. J. Futuyma. 1971. On the measurement of niche breadth and overlap. *Ecology* **52**: 567–576. (Chap. 10)

Condit, R., S. P. Hubbell, J. V. LaFrankie, R. Sukumar, N. Manokaran, R. Foster, and P. S. Ashton. 1996. Species-area and species-individual relationships for tropical trees: A comparison of three 50-ha plots. *Journal of Ecology* **84**: 549–562. (Chaps. 2, 6)

Connell, J. H. 1971. On the role of natural enemies in preventing competitive exclusion in some marine animals and in rain forest trees. Pp. 298–312 in P. J. den Boer and G. R. Gradwell, eds., *Dynamics of Populations*. Centre for Agricultural Publishing and Documentation, Wageningen, The Netherlands. (Chap. 1)

Connell, J. H. 1980. Diversity and the coevolution of competitors, or the ghost of competition past. *Oikos* **35**: 131–138. (Chap. 1)

Connor, E. F., and E. D. McCoy. 1979. The statistics and biology of the species-area relationship. *American Naturalist* **113**: 791–833. (Chap. 6)

Connor, E. F., and D. Simberloff. 1978. Species number and compositional similarity of the Galapagos flora and avifauna. *Ecological Monographs* **48**: 219–248. (Chap. 6)

Connor, E. F., D. McCoy, and B. J. Cosby. 1983. Model discrimination and expected slope values in species-area studies. *American Naturalist* **1122**: 789–796. (Chap. 6)

Cook, L. M., and G. S. Graham. 1996. Evenness and species number in some moth populations. *Biological Journal of the Linnean Society, London* **58**: 75–84. (Chap. 4)

Corbet, A. S. 1941. The distribution of butterflies in the Malay Peninsula. *Proceedings of the Royal Society, London, Series A* **16**: 101–116. (Chap. 2)

Croat, T. 1978. *The Flora of Barro Colorado Island.* Stanford University Press, Palo Alto, CA. (Chap. 6)

Crow, J. F., and M. Kimura. 1970. *An Introduction to Population Genetics Theory.* Harper and Row, New York. (Chap. 1)

Crowell, K. L. 1962. Reduced interspecific competition among the birds of Bermuda. *Ecology* **43**: 75–88. (Chap. 10)

Crowell, K. L. 1973. Experimental zoogeography: Introductions of mice to small islands. *American Naturalist* **107**: 535–558. (Chap. 10)

Curran, L. M., I. Caniago, G. D. Paoli, D. Astianti, M. Kusneti, M. Leighton, C. E. Nirarita, and N. Haeruman. 1999. Impact of El Niño and logging on canopy tree recruitment in Borneo. *Science* **286**: 2184–2188. (Chap. 3)

Dalling, J. W., M. D. Swaine, and N. C. Carwood. 1997. Community level soil seed bank dynamics in the 50 ha plot, on Barro Colorado Island, Panama. *Journal of Tropical Ecology* **13**: 659–680. (Chap. 10)

D'Arcy, W. E. 1987. *Flora of Panama: Checklist and Index.* Vols. 1–4. Missouri Botanical Garden, St. Louis. (Chap. 6)

Darwin, C. 1859. *On the Original of Species.* Jon Murry, London, UK. (Chap. 8)

Davis, M. B. 1976. Pleistocene biogeography of temperate deciduous forests. *Geoscience and Man* **13**: 13–26. (Chap. 1)

Davis, M. B. 1986. Climatic instability, time lags, and community disequilibrium. Pp. 269–284 in J. M. Diamond and T. J. Case, eds., *Community Ecology.* Harper and Row, New York. (Chap. 1)

Davis, M. B. 1989. Lags in vegetation response to greenhouse warming. *Climate Change* **15**: 75–82. (Chap. 1)

Davis, M. B. 1991. Research questions posed by the paleoecological record of global change. Pp. 385–395 in R. S. Bradley, ed., *Global Changes of the Past.* University Corporation for Atmospheric Research, Boulder, CO. (Chap. 1)

Delcourt, P. A., and H. R. Delcourt. 1987. *Long-Term Forest Dynamics of the Temperate Zone.* Springer-Verlag, New York. (Chap. 1)

Diamond, J. M. 1972. Biogeographic kinetics: Estimation of relaxation times for avifaunas of southwest Pacific islands. *Proceedings of the National Academy of Sciences, USA* **69**: 3199–3203. (Chap. 1)

Diamond, J. M. 1975. Assembly of species communities. Pp. 342–444 in M. L. Cody and J. M. Diamond, eds., *Ecology and Evolution of Communities.* Belknap Press of Harvard University Press, Cambridge, MA. (Chap. 1)

Diamond, J. M. 1984. Historic extinctions: A Rosetta stone for understanding prehistoric extinctions. Pp. 824–262 in P. S. Martin and R. G. Klein, eds., *Quaternary Extinctions: A Prehistoric Revolution.* University of Arizona Press, Tucson. (Chap. 1)

Diamond, J. M., and R. M. May. 1975. Island biogeography and the design of natural reserves. Pp. 228–252 in R. M. May, ed., *Theoretical Ecology.* Sinauer Associates, Sunderland, MA. (Chaps. 1, 7)

Durrett, R., and S. A. Levin. 1996. Spatial models for species-area curves. *Journal of Theoretical Biology* **179**: 119–127. (Chaps. 1, 6)

Eldridge, N., and S. J. Gould 1972. Punctuated equilibria: An alternative to phyletic gradualism. Pp. 82–115 in T.J.M. Schopf, ed. *Models in Paleobiology.* Freeman Cooper, San Francisco. (Chap. 8)

Engen, S. 1974. On species frequency models. *Biometrika.* **61**: 363–370. (Chap. 2)

Engen, S. 1978. *Stochastic Abundance Models with Emphasis on Biological Communities and Species Diversity.* Chapman and Hall, London. (Chaps. 2, 9)

Engen, S., and R. Lande. 1996. Population dynamic models generating the lognormal species abundance distribution. *Mathematical Biosciences* **132**: 169–183. (Chap. 3)

Erwin, T. L. 1982. Tropical forests: Their richness in Coleoptera and other arthropod species. *Coleopterist's Bulletin* **36**: 74–75. (Chap. 9)

Erwin, T. L. 1997. Biodiversity at its utmost: Tropical forest beetles. Pp. 27–40 in M. L. Reaka, D. E. Wilson, and E. O. Wilson, eds., *Biodiversity II.* Joseph Henry Press, Washington, DC. (Chap. 9)

Ewens, W. 1972. The sampling theory of selectively neutral alleles. *Theoretical Population Biology* **3**: 87–112. (Chaps. 5, 6, 8)

Fisher, R. A., A. S. Corbet, and C. B. Williams. 1943. The relation between the number of species and the number of individuals in a random sample of an animal population. *Journal of Animal Ecology* **12**: 42–58. (Chaps. 1, 2, 3, 5, 6, 9)

Gaston, K. J. 1994. *Rarity.* Chapman and Hall, London. (Chaps. 2, 6)

Geisel, A. 1955. Dr. Seuss' *On Beyond Zebra*. Random House, New York. (Chap. 1)

Gentry, A. H. 1982. Patterns of neotropical plant species diversity. *Evolutionary Biology* **15**: 1–85. (Chap. 10)

Gibbons, D. W., J. B. Reid, and R. A. Chapman. 1993. *The New Atlas of Breeding Birds in Britain and Ireland, 1988–1991*. T & A. D. Poyser, London. (Chaps. 2, 9)

Gilbert, F. S. 1980. The equilibrium theory of biogeography: Fact or fiction? *Journal of Biogeography* **7**: 209–215 (Chap. 6)

Gilpin, M. E., and I. Hanski. 1991. *Metapopulation Dynamics*. Academic Press, New York. (Chap. 1)

Gleason, H. A. 1922. On the relation between species and area. *Ecology* **3**: 158–162 (Chaps. 1, 6)

Gleason, H. A. 1926. The individualistic concept of the plant association. *Bulletin of the Torrey Botanical Club* **53**: 7–26. (Chap. 1)

Gleason, H. A. 1939. The individualistic concept of the plant association. *American Midland Naturalist* **21**: 92–110. (Chap. 1)

Gleick, J. 1987. *Chaos: Making a New Science*. Viking, New York. (Chaps. 1, 7.)

Goldberg, D. E., and P. A. Werner. 1983. Equivalence of competitors in plant communities: A null hypothesis and a field experimental test *American Journal of BotaNew York* **70**: 1098–1104. (Chap. 1)

Gotelli, N. J., and G. R. Graves. 1996. *Null Models in Ecology*. Smithsonian Institution Press, Washington, DC. (Chap. 1)

Gould, S. J., D. Raup, J. J. Sepkoski, Jr., T.J.M. Schopf, and D. S. Simberloff. 1977. The shape of evolution: A comparison of real and random clades. *Paleobiology* **3**: 23–40. (Chap. 8)

Grant, P. R. 1972. Convergent and divergent character displacement. *Biological Journal of the Linnean Society, London* **4**: 39–60. (Chap. 1)

Grant, P. R. 1975. The classical case of character displacement. *Evolutionary Biology* **8**: 237–337. (Chap. 1)

Grant, P. R. 1986. *Ecology and Evolution of Darwin's Finches*. Princeton University Press, Princeton, NJ. (Chap. 1)

Gray, J. S. 1979. Pollution-induced changes in populations. *Philosophical Transactions of the Royal Society, London, Series B* **286**: 545–561. (Chap. 9)

Gray, J. S. 1981. Detecting pollution induced changes in communities using the lognormal distribution of individuals among species. *Marine Pollution Bulletin* **12**: 173–176. (Chap. 9)

Gray, J. S. 1983. On the use and misuse of the lognormal plotting method for detection of effects of pollution. *Marine Ecology— Progress Series* **11**: 203–204. (Chap. 9)

Gray, J. S. 1987. Species abundance patterns. Pp. 53–67 in J.H.T. Gee and P. S. Giller, eds., *Organization of Communities, Past and Present.* Blackwell Scientific, Oxford, UK. (Chaps. 2, 9)

Gray, J. S., and F. B. Mirza. 1979. A possible method for the detection of pollution induced disturbance on marine benthic communities. *Marine Pollution Bulletin* **10**: 142–146. (Chap. 9)

Gregory, R. 1994. Species abundance patterns of British birds *Proceedings of the Royal Society, London, Series B* **257**: 299–301. (Chap. 2)

Grinnell, J. 1917. The niche relationships of the California thrasher. *Auk* **34**: 427–433. (Chap. 1)

Gullicksen, B., T. Haug, and O. K. Sandnes. 1980. Benthic macrofauna on new and old grounds at Jan Mayen. *Sarsia* **65**: 137–148. (Chap. 9)

Hågvar, S. 1994. Lognormal distribution of dominance as an indicator of stressed soil microarthropod communities? *Acta Zoologica Fennica* **195**: 71–80. (Chap. 9)

Hairston, N. G. 1959. Species abundance and community organization. *Ecology* **40**: 404–416. (Chap. 2)

Hairston, N. G., F. E. Smith, and L. B. Slobodkin. 1960. Community structure, population control, and competition. *American Naturalist* **94**: 421–425. (Chap. 1)

Hanski, I. A., and M. E. Gilpin. 1997. *Metapopulation Biology: Ecology, Genetics and Evolution.* Academic Press, San Diego, CA. (Chap. 1)

Hanski, I. A., and D. S. Simberloff 1997. The metapopulation approach, its history, conceptual domain, and application to conservation. Pp. 5–26 in I. A. Hanski and M. E. Gilpin, eds., *Metapopulation Biology: Ecology, Genetics, and Evolution.* Academic Press, San Diego, CA. (Chaps. 1, 7)

Harper, J. 1977. *Population Biology of Plants.* Academic Press, London. (Chap. 3)

Harrison, S. 1997. Spatial pattern formation in an insect host-parasitoid system. *Science* **2778**: 1619–1621. (Chap. 7)

Harte, J., and A. Kinzig. 1997. On the implications of species-area relationships for endemism, spatial turnover, and food web patterns. *Oikos* **80**: 417–327. (Chap. 7)

Harte, J., A. Kinzig, and J. Green. 1999. Self-similarity in the distribution and abundance of species. *Science* **284**: 334–336. (Chap. 6)

LITERATURE CITED

Hartnol, R. G., M. T. Burrows, and F. M. Ellard. 1985. Species-abundance distributions: Arbiters or artifacts? Pp. 381–390 in P. G. Gibbs, ed., *Proceedings of the 19th European Marine Biology Symposium.* Cambridge University Press, Cambridge, UK. (Chap. 9)

Harvey, P. H., and S. Nee. 1994. Comparing real with expected patterns from molecular phylogenies. Pp. 219–231 in P. Eggleton and R. I. Vane-Wright, eds., *Phylogenetics and Ecology.* Acadamic Press, London. (Chap. 8)

Harvey, P. H., E. C. Holmes, A. Ø. Mooers, and S. Nee. 1994a. Inferring evolutionary processes from molecular phylogenies. Pp. 314–333 in R. W. Scotland, D. J. Siebert, and D. M. Williams, eds. *Models in Phylogeny Reconstruction.* Systematics Association, special vol. no. 52. Clarendon Press, Oxford. (Chap. 8)

Harvey, P. H., R. M. May, and S. Nee. 1994b. Phylogenies without fossils. *Evolution* **48**: 523–529. (Chap. 8)

Harvey, P. H., A. J. Leigh Brown, J. Maynard Smith, and S. Nee, eds. 1996. *New Uses for New Phylogenies.* Oxford University Press, Oxford. (Chap. 8)

Hey, J. 1991. The structure of genealogies and the distribution of fixed differences between DNA sequence samples from natural populations. *Genetics* **128**: 831–840. (Chap. 8)

Hey, J. 1992. Using phylogenetic trees to study speciation and extinction. *Evolution* **46**: 627–640. (Chap. 8)

Hey, J., H. Hilton, N. Leahy, and R.-L. Wang. 1993. Testing models of speciation and extinction with phylogenetic trees of extant taxa. Pp. 70–90 in M. L. McKinney and J. A. Drake, eds., *Biodiversity Dynamics: Turnover of Populations, Taxa, and Communities.* Columbia University Press, New York. (Chap. 8)

Hicks, G.R.F. 1980. Structure of phytal harpacticoid copepod assemblages and the influence of habitat complexity and trubidity. *Journal of Experimental Marine Biology and Ecology* **44**: 157–192. (Chap. 9)

Holley, R. A., and T. M. Liggest. 1975. Ergodic theorems for weakly interacting systems and the voter model. *Annals of Probability* **3**: 643–663. (Chap. 6)

Holmes, R. T., T. W. Sherry, and F. W. Sturgis. 1986. Bird community dynamics in a temperate deciduous forest: Long-term trends at Hubbard Brook. *Ecological Monographs* **56**: 201–220. (Chap. 9)

Holt, R. D. 1977. Predation, apparent competition, and the structure of prey communities. *Theoretical Population Biology* **12**: 197–229. (Chap. 1)

Hsu, S. B., S. P. Hubbell, and P. Waltman. 1977. A mathematical theory for single-nutrient competition in continuous cultures of microorganisms. *Society for Industrial and Applied Mathematics, Journal of Applied Mathematics* **32**: 266–383. (Chap. 2)

Hubbell, S. P. 1979. Tree dispersion, abundance and diversity in a tropical dry forest. *Science* **203**: 1299–1309. (Chaps. 1, 3, 5)

Hubbell, S. P. 1995. Toward a theory of biodiversity and biogeography on continuous landscapes. Pp. 171–199 in G. R. Carmichael, G. E. Folk, and J. E. Schnoor, eds., *Preparing for Global Change: A Midwestern Perspective.* Academic Publishing, Amsterdam. (Chap. 1)

Hubbell, S. P. 1997. A unified theory of biogeography and relative species abundance and its application to tropical rain forests and coral reefs. *Coral Reefs* **16** (suppl.): S9–S21. (Chaps. 1, 5)

Hubbell, S. P., and R. B. Foster. 1983. Diversity of canopy trees in a neotropical forest and implications for the conservation of tropical trees. Pp. 25–41 in S. J. Sutton, T. C. Whitmore, and A. C. Chadwick, eds., *Tropical Rain Forest: Ecology and Management.* Blackwell, Oxford. (Chap. 3)

Hubbell, S. P., and R. B. Foster. 1986a. Commonness and rarity in a neotropical forest: Implications for tropical tree conservation. Pp. 205–231 in M. Soulé, ed., *Conservation Biology: Science of Scarcity and Diversity.* Sinauer Associates, Sunderland, MA. (Chap. 2)

Hubbell, S. P., and R. B. Foster. 1986b. Biology, chance and history and the structure of tropical rain forest tree communities. Pp. 314–329 in J. M. Diamond and T. J. Case, eds., *Community Ecology.* Harper and Row, New York. (Chaps. 4, 10)

Hubbell, S. P., and R. B. Foster. 1990. Structure, dynamics and equilibrium status of old-growth forest on Barro Colorado Island. Pp. 522–541 in A. Gentry, ed., *Four Neotropical Forests.* Yale University Press, New Haven, CT. (Chap. 3)

Hughes, R. G. 1984. A model of the structure and dynamics of benthic marine invertebrate communities. *Marine Ecology—Progress Series.* **15**: 1–11. (Chaps. 3, 9)

Hughes, R. G. 1986. Theories and models of species abundance. *American Naturalist* **128**: 879–899. (Chaps. 3, 9)

Huntley, B., and T. Webb III. 1989. Migration: Species' responses to climatic variation caused by changes in the earth's orbit. *Journal of Biogeography* **16**: 5–19. (Chap. 1)

Hurtt, G. C., and S. W. Pacala. 1995. The consequences of recruitment limitation: Reconciling chance, history, and competitive differences between plants. *Journal of Theoretical Biology* **176**: 1–12. (Chaps. 3, 7, 10)

Huston, M. A. 1994. *Biological Diversity: The Coexistence of Species on Changing Landscapes.* Cambridge University Press, Cambridge, UK. (Chap. 1)

Jablonski, D. 1995. Extinctions in the fossil record. Pp. 25–44 in J. H. Lawton and R. M. May, eds., *Extinction Rates.* Oxford University Press, Oxford. (Chap. 10)

Jackson, J.B.C. 1995. Constancy and change in the life of the sea. Pp. 45–54 in J. H. Lawton and R. M. May, eds., *Extinction Rates.* Oxford University Press, Oxford. (Chap. 10)

Jackson, J.B.C., A. F. Budd, and J. M. Pandolfi. 1996. The shifting balance of natural communities? Pp. 89–127 in D. Jablonski, D. H. Erwin, and J. H. Lipps, eds., *Evolutionary Paleobiology.* University of Chicago Press, Chicago. (Chaps. 1, 3, 10)

Janzen, D. H. 1970. Herbivores and the number of tree species in tropical forests. *American Naturalist* **104**: 501–528. (Chap. 1)

Janzen, D. H. 1974. Tropical blackwater rivers, animals, and mast fruiting of the Dipterocarpaceae. *Biotropica* **6**: 69–113. (Chap. 3)

Janzen, D. H. 1997. Wildlife biodiversity management in the tropics. Pp. 411–431 in M. L. Reaka, D. E. Wilson, and E. O. Wilson, eds., *Biodiversity II.* Joseph Henry Press, Washington, DC. (Chap. 9)

Janzen, D. H., and W. Hallwachs. 1994. *All Taxa Biodiversity Inventory (ATBI) of Terrestrial Systems.* Report of an NSF Workshop, April 1993, Philadelphia. (Chap. 9)

Johnson, M. P., and P. H. Raven. 1973. Species number and endemism: The Galapagos archipelago revisited. *Science* **179**: 893–895. (Chap. 6)

Kalko, E.K.V., C. O. Handley, and D. Handley. 1996. Organization, diversity, and long-term dynamics of a neotropical bat community. Pp. 503–553 in M. L. Cody and J. A. Smallwood, eds., *Long-Term Studies of Vertebrate Communities.* Academic Press, San Diego, CA. (Chap. 9)

Kareiva, P., and U. Wennegren. 1995. Connecting landscape patterns to ecosystem and population processes. *Nature* **373**: 299–302. (Chap. 7)

Karlin, S., and J. McGregor. 1972. Addendum to a paper of W. Ewens. *Theoretical Population Biology* **3**: 113–116. (Chap. 5)

Keiher, E., and P. Keddy, eds. 1999. *Ecological Assembly Rules: Perspectives, Advances, Retreats.* Cambridge University Press, Cambridge, UK. (Chap. 1)

Kemeny, J. C., and J. L. Snell. 1960. *Finite Markov Chains.* Van Nostand Reinhold, New York. (Chap. 4)

Kempton, R. A., and L. R. Taylor. 1974. Log-series and log-normal parameters as diversity discriminants for the Lepidoptera. *Journal of Animal Ecology* **43**: 381–390. (Chap. 9)

Kendall, D. G. 1948. On the generalized birth-and-death process. *Annals of Mathematical Statistics* **19**: 1–15. (Chap. 8)

Kevan, P. G., C. F. Greco, and S. Belaoussoff. 1997. Log-normality of biodiversity and abundance in diagnosis and measuring of ecosystem health: Pesticide stress on pollinators on blueberry heaths. *Journal of Applied Ecology* **34**: 1122–1136. (Chap. 9)

Kevles, D. 1971. *The Physicists.* Belknap Press of Harvard University Press, Cambridge, MA. (Chap. 1)

Kilburn, P. D. 1966. Analysis of the species-area relation. *Ecology* **47**: 831–843. (Chap. 6)

King, C. E. 1964. Relative abundance of species and MacArthur's model. *Ecology* **45**: 716–727. (Chap. 2)

Klinge, J., and R. M. Zink. 1997. The importance of recent Ice Ages in speciation: A failed paradigm. *Science* **177**: 166–169. (Chap. 8)

Kohn, A. J. 1969. The ecology of the genus *Conus* in Hawaii. *Ecological Monographs* **29**: 47–90. (Chap. 2).

Kolasa, J., and S.T.A. Pickett, eds. 1991. *Ecological Heterogeneity.* Ecological Studies no. 86. Springer-Verlag, New York. (Chap. 1)

Krebs, J. R., and N. B. Davies. 1978. *Behavioral Ecology: An Evolutionary Approach.* Sinauer Associates, Sunderland, MA. (Chap. 1)

Lambshead, J., and H. M. Platt. 1985. Structural patterns of marine benthic assemblages and their relationships with empirical statistical models. Pp. 371–380 in P. E. Gibbs, ed., *Proceedings of the 19th European Marine Biology Symposium, Plymouth, 1984.* Cambridge University Press, Cambridge, UK. (Chap. 9)

Lande, R. 1988. Genetics and demography in biological conservation. *Science* **241**: 1455–1460. (Chaps. 8, 9)

Lande, R. 1993. Risks of population extinction from demographic and environmental stochasticity and random catastrophes. *American Naturalist* **142**: 911–927. (Chaps. 4, 8, 9).

LaRoi, G. H. 1967. Ecological studies of the boreal spruce-fir forests of the North American taiga. I. Analysis of the vascular flora. *Ecological Monographs* **37**: 239–253. (Chap. 7)

LaRoi, G. H., and M. H. Stringer. 1976. Ecological studies on the boreal spruce-fir forests of the North American taiga. II. Analysis of the bryophyte flora. *Canadian Journal of Botany* **54**: 619–643. (Chap. 7)

Lawton, J. H. 1990. Species richness and population dynamics: Patterns in body size: abundance space. *Philosophical Transactions of the Royal Society, London, Series B* **330**: 283–291. (Chap. 1)

Lawton, J. H., S. Nee, A. J. Letcher, and P. H. Harvey. 1994. Animal distributions: Patterns and Processes. Pp. 41–58 in P. J. Edwards, R. M. May, and N. R. Webb, eds., *Large-Scale Ecology and Conservation Biology*. Blackwell, Oxford. (Chap. 1)

Levin, S. A. 1970. Community equilibria and stability, and an extension of the competitive exclusion principle. *American Naturalist* **104**: 413–423. (Chap. 1)

Levins, R. 1968. *Evolution in Changing Environments*. Monographs in Population Biology. Princeton University Press, Princeton, NJ. (Chaps. 2, 10)

Levins, R. 1969. Some demographic and genetic consequences of environmental heterogeneity for biological control. *Bulletin of the Entomological Society of America* **15**: 237–240. (Chaps. 1, 10)

Levins, R. 1970. Extinction. Pp. 75–107 in M. Gerstenhaber, ed., *Some Mathematical Problems in Biology*, American Mathematical Society, Providence, RI. (Chap. 8)

Levins, R. 1975. Evolution in communities near equilibrium. Pp. 16–50 in M. L. Cody and J. M. Diamond, eds., *Ecology and Evolution of Communities*. Belknap Press of Harvard University Press, Cambridge, MA. (Chap. 1)

Levinton, J. S. 1979. A theory of diversity equilibrium and morphological evolution. *Science* **204**: 335–336. (Chap. 8)

Lewontin, R. C. 1974. *The Genetic Basis of Evolutionary Change*. Columbia University Press, New York. (Chap. 1)

MacArthur, R. H. 1957. On the relative abundance of species. *Proceedings of the National Academy of Sciences, USA* **43**: 293. (Chaps. 1, 2)

MacArthur, R. H. 1960. On the relative abundance of species. *American Naturalist* **94**: 25–36. (Chaps. 1, 2)

MacArthur, R. H. 1970. Species packing and competitive equilibrium for many species. *Theoretical Population Biology* **1**: 1–11. (Chap. 1)

MacArthur, R. H. 1972. *Geographical Ecology*. Harper and Row, New York. (Chap. 1)

MacArthur, R. H., and E. O. Wilson. 1963. An equilibrium theory of insular zoogeography. *Evolution* **17**: 373–387. (Chap. 1)

MacArthur, R. H., and E. O. Wilson. 1967. *The Theory of Island Biogeography*. Monographs in Population Biology. Princeton University Press, Princeton, NJ. (Chaps. 1, 2, 3, 5, 6, 7, 10)

Magurran, A. E. 1988. *Ecological Diversity and Its Measurement*. Princeton University Press, Princeton, NJ. (Chaps. 2, 9)

Mandelbrot, B. B. 1982. *The Fractal Geometry of Nature*. W. H. Freeman, New York. (Chap. 7)

Manly, B.F.J. 1991. *Randomization and Monte Carlo Methods in Biology.* Chapman and Hall, New York. (Chap. 7)

Manokaran, N., J. V. LaFrankie, K. M. Kochummen, E. S. Quah, J. E. Klahn, P. S. Ashton, and S. P. Hubbell. 1992. *Stand Tables and Species Distributions in the Fifty Hectare Plot at Pasoh Forest Reserve.* Forest Research Institute—Malaysia, Kuala Lumpur. (Chaps. 3, 5)

May, R. M. 1973. *Stability and Complexity in Model Ecosystems.* Monographs in Population Biology. Princeton University Press, Princeton, NJ. (Chap. 1)

May, R. M. 1975. Patterns of species abundance and diversity. Pp. 81–120 in M. L. Cody and J. M. Diamond, eds., *Ecology and Evolution of Communities.* Belknap Press of Harvard University Press, Cambridge, MA. (Chaps. 1, 2, 3, 6, 9)

May, R. M. 1988. How many species are there on Earth? *Science* **241**: 1441–1449. (Chap. 9)

May, R. M., and J. H. Lawton. 1995. Assessing extinction rates. Chapter 1 in J. H. Lawton and R. M. May, eds., *Extinction Rates.* Oxford University Press, Oxford. (Chap. 1)

Maynard Smith, J. 1968. *Mathematical Ideas in Biology.* Cambridge University Press, London. (Chap. 1)

Maynard Smith, J. 1974. *Models in Ecology.* Cambridge University Press, London. (Chap. 1)

Mayr, E. 1963. *Animal Species and Evolution.* Belknap Press of Harvard University Press, Cambridge, MA. (Chaps. 5, 6, 8, 10)

McGowan, J. A., and P. W. Walker. 1993. Pelagic diversity patterns. Pp. 203–214 in R. E. Ricklefs and D. Schluter, eds., *Species Diversity in Ecological Communities.* University of Chicago Press, Chicago. (Chap. 9)

McGuinness, K. A. 1984. Equations and explanations in the study of species-area curves. *Biological Reviews* **59**: 423–440. (Chap. 6)

McKinney, M. L., and J. A. Drake, eds. 1993. *Biodiversity Dynamics: Turnover of Populations, Taxa, and Communities.* Columbia University Press, New York. (Chap. 8)

Milne, B. T. 1997. Applications of fractal geometry in wildlife biology. Pp. 32–69 in J. A. Bisonette, ed., *Wildlife and Landscape Ecology.* Springer-Verlag, New York (Chap. 7)

Mizra, F. D., and J. S. Gray. 1981. The fauna of benthic sediments from the organically enriched Oslofjord, Norway. *Journal of Experimental Marine Biology and Ecology* **54**: 181–207. (Chap. 9)

LITERATURE CITED

Monroe, E. G. 1948. The geographical distribution of butterflies in the West Indies. Ph.D. diss., Cornell University, Ithaca, New York. (Chap. 1)

Monroe, E. G. 1953. The size of island faunas. Pp. 52–53 in *Proceedings of the Seventh Pacific Science Congress of the Pacific Science Association*, vol. 4, *Zoology*. Whitcome and Tombs, Aukland, New Zealand. (Chap. 1)

Motomura, I. 1932. A statistical treatment of associations. (In Japanese.) *Zoological Magazine, Tokyo* **44**: 379–383. (Chaps. 1,2,5,10)

Muller-Dombois, D., and H. Ellenberg. 1974. *Aims and Methods of Vegetation Ecolgoy*. John Wiley and Sons, New York. (Chap. 7)

Myers, N. 1988. Tropical forests and their species: Going, going, ...? Pp. 28–35 in E. O. Wilson, ed., *Biodiversity*. National Academy Press, Washington, DC. (Chap. 1)

Naeem, S., and R. K. Colwell. 1991. Ecological consequences of heterogeneity in consumable resources. Pp. 224–255 in J. Kolasa and S.T.A. Pickett, eds., *Ecological Heterogeneity*. Ecological Studies no. 86. Springer-Verlag, New York. (Chap. 1)

Nason, J. D., E. A. Herre, and J. L. Hamrick. 1996. Paternity analysis of the breeding structure of strangler fig populations: Evidence for substantial long-distance wasp dispersal. *Journal of Biogeography* **23**: 501–512. (Chap. 6)

National Academy of Sciences. 1997. *Genetics and the Origin of Species: From Darwin to Molecular Biology 60 Years after Dobzhansky*. NAS Colloquium, Proceedings of the National Academy of Sciences, USA. (Chap. 8)

Nee, S., A. Ø. Mooers, and P. H. Harvey. 1992. The tempo and mode of evolutionary biology revealed from molecular phylogenies. *Proceedings of the National Academy of Sciences, USA* **89**: 8322–8326. (Chap. 8)

Nee, S., R. M. May, and P. H. Harvey. 1994. The reconstructed evolutionary process. *Philosophical Transactions of the Royal Society, London, Series B* **344**: 305–311. (Chap. 8)

Nee, S., E. C. Holmes, R. M. May, and P. H. Harvey. 1995. Estimating extinction rates from molecular phylogenies. Pp. 164–182 in J. H. Lawton and R. M. May, eds., *Extinction Rates*. Oxford University Press, Oxford. (Chap. 8)

Nekola, J. C., and P. S. White. 1999. The distance decay of similarity in biogeography and ecology. *Journal of Biogeography* **26**: 867–878. (Chap. 7)

Oksanen, L. 1988. Ecosystem organization: Mutualism and cybernetics or plain Darwinian struggle for Existence? *American Naturalist* **131**: 424–444. (Chap. 1)

Ortner, P. B., L. C. Hill, and S. R. Cummings. 1982. Variation in copepod species assemblage distributions. The utility of a log-normal approach. *Marine Pollution Bulletin* **13**: 195–197. (Chap. 9)

Otte, J., and J. A. Endler, eds. 1989. *Speciation and Its Consequences.* Sinauer Associates, Sunderland, MA. (Chap. 8)

Overpeck, J. T., R. S. Webb, and T. Webb III. 1992. Mapping eastern North American vegetation change of the last 18 Ka: vNo-analogs and the future. *Geology* **20**: 1071–1074. (Chap. 1)

Pagels, H. R. 1982. *The Cosmic Code.* Simon and Schuster, New York. (Chap. 1)

Paine, R. T. 1966. Food web complexity and species diversity. *American Naturalist* **100**: 65–75. (Chap. 1)

Palmer, M. W., and P. S. White. 1994. Scale dependence and the species-area relationship. *American Naturalist* **144**: 717–740. (Chaps. 6, 7)

Pandolfi, J. M. 1996. Limited membership in Pleistocene reef coral assemblages from the Huon Peninsula, Papua New Guinea: Constancy during global change. *Paleobiology* **22**: 152–176. (Chap. 10)

Pandolfi, J. M., and P. R. Minchin. 1995. A comparison of taxonomic composition and diversity between reef coral life and death assemblages in Madang Lagoon, Papua New Guinea. *Palaeogeography, Palaeoclimatology, Palaeoecology* **119**: 321–341. (Chap. 10)

Patrick, R. 1968. The structure of diatom communities in similar ecological conditions. *American Naturalist* **102**: 173–183. (Chap. 9)

Patrick, R. 1972. Benthic communities in streams. *Transactions of the Connecticut Academy of Arts and Sciences* **44**: 379–383. (Chap. 9)

Patrick, R., M. Hohn, and J. Wallace. 1954. A new method of determining the pattern of the diatom flora. Academy of Natural Sciences, Philadelphia, *Notulae Naturae* **259**. (Chap. 9)

Patzkowsky, M. F., and S. M. Holland. 1997. Patterns of turnover in Middle to Upper Ordovician brachiopods of the eastern United States: A test of coordinated stasis. *Paleobiology* **23**: 420–443. (Chaps. 1, 8)

Pielou, E. C. 1966. The measurement of diversity in different types of biological collections. *Journal of Theoretical Biology* **13**: 131–144. (Chap. 2)

Pielou, E. C. 1975. *Ecological Diversity.* Wiley, New York. (Chaps. 2, 9)

Pielou, E. C. 1977. *An Introduction to Mathematical Ecology.* 2nd ed. Wiley Interscience, New York. (Chap. 1)

Pielou, E. C. 1979. *Biogeography*. Wiley Interscience, New York.

Pimm, S. L. 1982. *Food Webs*. Chapman and Hall, London. (Chap. 1)

Pimm, S. L. 1991. *The Balance of Nature? Ecological Issues in the Conservation of Species and Communities*. University of Chicago Press, Chicago. (Chap. 1)

Pontin, A. J. 1982. *Competition and Coexistence of Species*. Pitman, London. (Chap. 1)

Preston, F. W. 1948. The commonness, and rarity, of species. *Ecology* **29**: 254–283. (Chaps. 1, 2, 3, 5, 8, 9)

Preston, F. W. 1960. Time and space variation of species. *Ecology* **41**: 611–627. (Chap. 6)

Preston, F. W. 1962. The canonical distribution of commonness and rarity. *Ecology* **43**: 185–215, 410–432. (Chaps. 1, 2, 3, 8, 9)

Preston, F. W. 1980. Noncanonical distributions of commonness and rarity. *Ecology* **61**: 88–93. (Chaps. 2, 3, 9)

Pulliam, H. R. 1975. Coexistence of sparrows: A test of competition theory. *Science* **189**: 474–476. (Chap. 1)

Pulliam,. H. R. 1983. Ecological community theory and the coexistence of sparrows. *Ecology* **64**: 45–52. (Chap. 1)

Pulliam, H. R. 1988. Sources, sinks, and population regulation. *American Naturalist* **132**: 652–661. (Chap. 10)

Quinn, J. F., and A. Hastings. 1987. Extinction in subdivided populations. *Conservation Biology* **1**: 198–208. (Chap. 1)

Rabinowitz, D., S. Cairns, and T. Dillon. 1986. Seven forms of rarity and their frequency in the flora of the British Isles. Pp. 182–204 in M. E. Soulé, ed., *Conservation Biology: The Science of Scarcity and Diversity*. Sinauer Associates, Sunderland, MA. (Chap. 2)

Rao, C. R. 1971. Some comments on the logarithmic series distribution in the analysis of insect trap data. Pp. 131–142 in G. P. Patil, E. C. Pielou, and W. E. Waters, eds., *Statistical Ecology*, vol. 1. Pennsylvania State Univerisity Press, University Park, PA. (Chap. 9)

Raup, D. M. 1978. Cohort analysis of genetic survivorship. *Paleobiology* **4**: 1–15. (Chap. 8)

Raup, D. M. 1985. Mathematical models of cladogenesis. *Paleobiology* **11**: 47–52. (Chap. 8)

Raup, D. M. 1991. A kill curve for Phanerozoic marine species. *Paleobiology* **17**: 37–48. (Chaps. 8, 10)

Raup, D. M., S. J. Gould, T.J.M. Schopf, and D. S. Simberloff. 1973. Stochastic models of phylogeny and the evolution of diversity. *Journal of Geology* **81**: 525–542. (Chap. 8)

Reich, P. B., M. B. Walters, and D. S. Ellsworth. 1997. From tropics to tundra: Global convergence in plant functioning. *Proceedings of the National Academy of Sciences, USA* **94**: 3730–3734. (Chap. 10)

Reich, P. B., M. B. Walters, and D. S. Ellsworth, J. M. Vose, J. C. Volin, C. Gresham, and W. D. Bowman. 1998. Relationship of leaf dark respiration to leaf nitrogen, specific leaf area, and leaf life span: A test across biomes and functional groups *Oecologia* **114**: 471–482. (Chapt. 10)

Reich, P. B., M. B. Walters, and D. S. Ellsworth, J. M. Vose, J. C. Volin, and W. D. Bowman. 1999. Generality of leaf trait relationships: A test across six biomes. *Ecology* **80**: 1955–1969. (Chapt. 10)

Ricklefs, R. E., and D. Schluter, eds. 1993. *Species Diversity in Ecological Communities.* University of Chicago Press, Chicago. (Chap. 1)

Richter-Dyn, N., and S. S. Goel. 1972. On the extinction of a colonizing species. *Theoretical Population Biology* **3**: 406–433. (Chaps. 8, 9).

Ritchie, M. E. 1997. Populations in a landscape context: Sources, sinks and metapopulations. Pp. 160–184 in J. A. Bissonette, ed., *Wildlife and Landscape Ecology.* Springer-Verlag, New York. (Chap. 7)

Ritchie, M. E., and H. Olff. 1999. Spatial scaling laws yield a synthetic theory of biodiversity. *Nature* **400**: 557–560. (Preface, Chaps. 7, 10)

Rose, M. R. 1987. *Quantitative Ecological Theory.* Johns Hopkins University Press, Baltimore, MD. (Chaps. 1, 2)

Rosenzweig, M. L. 1975. On continental steady-states of species diversity. Pp. 121–140 in M. L. Cody and J. M. Diamond, eds., *The Ecology and Evolution of Communities.* Belknap Press of Harvard University Press, Cambridge, MA. (Chap. 8)

Rosenzweig, M. L. 1978. Competitive speciation. *Biological Journal of the Linnean Society, London* **10**: 275–289. (Chap. 5)

Rosenzweig, M. L. 1995. *Species Diversity in Space and Time.* Cambridge University Press, Cambridge, UK. (Chaps. 1, 6, 8)

Roughgarden, J. 1972. Evolution of niche width. *American Naturalist* **106**: 683–718. (Chap. 10)

Routledge, R. D. 1980. The form of species abundance distributions. *Journal of Theoretical Biology* **82**: 547–558. (Chaps. 2, 9)

Sale, P. F. 1977. Maintenance of high diversity in coral reef fish communities. *American Naturalist* **111**: 337–359. (Chap. 3)

Sale, P. F. 1980. The ecology of fishes on coral reefs. *Oceanography and Marine Biology Annual Reviews* **18**: 367–421. (Chap. 3)

Sanders, H. L. 1968. Marine benthic diversity: A comparative study. *American Naturalist* **102**: 243–282. (Chap. 5)

Sanders, H. L. 1969. Benthic marine diversity and the stability-time hypothesis. *Brookhaven Symposia in Biology* **22**: 71–81. (Chaps. 5, 6)

Scotland, R. W., D. J. Siebert, and D. M. Williams, eds. 1994. *Models in Phylogeny Reconstruction.* Systematics Association, special vol. no. 52. Clarendon Press, Oxford. (Chap. 8)

Shmida, A., and S. Ellner. 1984. Coexistence of plant species with similar niches. *Vegetatio* **58**: 29–55. (Chaps. 1, 7)

Shmida, A., and M. V. Wilson. 1985. Biological determinants of species diversity. *Journal of Biogeography* **12**: 1–20. (Chap. 6)

Sibley, C. G., and J. E. Ahlquist. 1990. *Phlogeny and Classsification of Birds.* Yale University Press, New Haven, CT. (Chap. 8)

Silvertown J. S. Holtier, J. Johnson, and P. Dale. 1992. Cellular automaton models of interspecific competition for space— the effect of pattern on process. *Journal of Ecology* **80**: 527–534. (Chap. 6)

Simberloff, D. S. 1969. Experimental zoogeography of islands: A model for insular colonization. *Ecology* **50**: 296–314. (Chap. 1)

Singh, R. S. 1989. Patterns of species divergence and genetic theories of speciation. Pp. 231–265 in K. K. Wöhrman and S. K. Jain, eds., *Population Biology.* Springer-Verlag, Berlin. (Chap. 5)

Slatkin, M. 1977. Gene flow and genetic drift in a species subject to frequent local extinction. *Theoretical population Biology* **12**: 3253–3262. (Chap. 6)

Slatkin, M. 1980. The distribution of mutant alleles in a subdivided population. *Genetics* **95**: 503–524. (Chap. 6)

Slobodkin, L. B. 1996. Islands of peril and pleasure. *Nature* **361**: 205–206. (Chap. 1)

Slobodkin, L., and H. L. Sanders. 1969. On the contribution of environmental predictability to species diversity. Pp.82–95 in *Diversity and Stability in Ecological Systems.* Brookhaven Symposia in Biology, no. 22. Brookhaven National Laboratory, Upton, New York. (Chap. 5)

Slobodkin, L., F. E. Smith, and N. G. Hairston, Sr. 1967. Regulation in terrestrial ecosystems and the implied balance of nature. *American Naturalist* **101**: 109–124. (Chap. 1)

Slocumb, J., B. Stauffer, and K. L. Dickson. 1977. On fitting the truncated log-normal distribution to species-abundance data using maximum likelihood estimation. *Ecology* **56**: 693–696. (Chaps. 2, 9)

Soulé, M. E., ed. 1986. *Conservation Biology: Science of Scarcity and Diversity.* Sinauer Associates, Sunderland, MA. (Chap. 2)

Soulsby P. G., D. Lowthion, and M. Houston. 1982. Effects of macroalgal mats on the ecology of intertidal mudflats. *Marine Pollution Bulletin* **13**: 162–166. (Chap. 9)

Stanley, S. M. 1979. *Macroevolution.* W. H. Freeman, San Francisco. (Chap. 8)

Stebbins, G. L. 1950. *Variation and Evolution in Plants.* Columbia University Press, New York. (Chap. 5)

Stork, N. E. 1997. Measuring global biodiversity and its decline. Pp. 41–68 in M. L. Reaka, D. E. Wilson, and E. O. Wilson, eds., *Biodiversity II.* Joseph Henry Press, Washington, DC. (Chaps. 9, 10)

Strong, D. R., D. Simberloff, L. G. Abele, and A. B. Thistle, eds. 1984. *Ecological Communities: Conceptual Issues and the Evidence.* Princeton University Press, Princeton, NJ. (Chap. 1)

Sugihara, G. 1980. Minimal community structure: An explanation of species abundance patterns. *American Naturalist* **116**: 770–787. (Chaps. 1, 2, 3, 9)

Sugihara, G. 1981. $S = cA^z$, $z = 1/4$: A reply to Connor and McCoy. *American Naturalist* **117**: 790–793. (Chap. 6)

Tansley, A. G. 1935. The use and abuse of vegetational concepts and terms. *Ecology* **16**: 284–307. (Chap. 1)

Taylor, L. R., R. A. Kempton, and I. P. Woiwod. 1976. Diversity statistics and the log-series model. *Journal of Animal Ecology* **45**: 255–271. (Chap. 9)

Templeton, A. R. 1981. Mechanisms of speciation—a population genetics approach. *Annual Review of Ecology and Systematics* **12**: 23–48. (Chap. 5)

Terborgh, J. 1973. On the notion of favorableness in plant ecology. *American Naturalist* **107**: 487–501. (Chap. 5)

Terborgh, J. 1974. Preservation of natural diversity: The problem of extinction-prone species. *Bioscience* **24**: 515–722. (Chap. 1)

Terborgh, J., and B. Winter. 1980. Some causes of extinction. Pp. 119–133 in M. E. Soulé and B. A. Wilcox, eds., *Conservation Biology: An Evolutionary-Ecological Perspective.* Sinauer Associates, Sunderland, MA. (Chap. 1)

Terborgh, J., S. K. Robinson, T. A. Parker III, C. A. Munn, and N. Pieront. 1990. Structure and organization of an Amazonian forest bird community. *Ecological Monographs* **60**: 213–238. (Chaps. 9, 10).

Terborgh, J., R. B. Foster, and V. Percy Nuñez. 1996. Tropical tree communities: A test of the nonequilibrium hypothesis. *Ecology* **77**: 561–567. (Chap. 10)

LITERATURE CITED

Thompson, G. B., and P.K.S. Shin. 1983. Sewage pollution and the infaunal benthos of Victoria Harbour, Hong Kong. *Journal of Experimental Marine Biology and Ecology* **67**: 279–299. (Chap. 9)

Tilman, D. 1982. *Resource Competition and Community Structure.* Monographs in Population Biology. Princeton University Press, Princeton, NJ. (Chaps. 1, 2)

Tilman, D. 1987. The importance of the mechanisms of interspecific competition. *American Naturalist* **129**: 769–774. (Chap. 1)

Tilman, D. 1988. *Plant Strategies and the Dynamics and Structure of Plant Communities.* Monographs in Population Biology. Princeton University Press, Princeton, NJ. (Chap. 10)

Tilman, D. 1989. Ecological experimentation: Strengths and conceptual problems. Pp. 136-157 in G. E. Likens, ed., *Long-Term Studies in Ecology: Approaches and Alternatives.* Springer-Verlag, New York. (Chap. 1)

Tilman, D. 1994. Competition and biodiversity in spatially structured habitats. *Ecology* **75**: 2–16. (Chaps. 3, 7, 10)

Tilman, D. 2000. Global environmental impacts of agricultural expansions: The need for sustainable and efficient practices. *Proceedings Nat. Acad. Sci., USA* **96**: 5995–2000. (Preface)

Tilman, D., and S. Pacala. 1993. The maintenance of species richness in plant communities. Pp. 13–25 in R. E. Ricklefs and D. Schluter, eds., *Species Diversity in Ecological Communities.* University of Chicago Press, Chicago. (Chap. 10)

Tilman, D., R. M. May, C. L. Lehman, and M. A. Nowak. 1994. Habitat destruction and the extinction debt. *Nature* **371**: 365–366. (Chap. 7)

Tilman, D., C. L. Lehman, and Yin Chengjun. 1997. Habitat destruction, dispersal, and deterministic extinction in competitive communities. *Proceedings of the National Academy of Sciences, USA* **94**: 1857–1861. (Chap. 7)

Tokeshi, M. 1993. Species abundance patterns and community structure. *Advances in Ecological Research* **24**: 111–186. (Chaps. 1, 2, 9)

Tokeshi, M. 1997. Species coexistence and abundance: Patterns and processes. Pp. 9–20 in A. Takuay, S. A. Levin, and M. Higashi, eds., *Biodiversity: An Ecological Perspective.* Springer-verlag, New York. (Chaps. 2, 9)

Tokeshi, M., 1999. *Species Coexistence: Ecological and Evolutionary Perspectives.* Blackwell, Oxford. (Chaps. 1, 9)

Turelli, M. 1978a. A reexamination of stability in randomly varying versus deterministic environments with comments on the stochastic theory of limiting similarity. *Theoretical Population Biology* **13**: 244–267. (Chap. 3)

Turelli, M. 1978b. Does environmental variability limit niche overlap? *Proceedings of the National Academy of Sciences, USA* **75**: 5085–5089. (Chap. 3)

Turelli, M., and M. E. Gilpin. 1980. Conditions for the existence of stationary densities for some two-dimensional diffusion processes with application to population biology. *Theoretical Population Biology* **17**: 167–189. (Chap. 3)

Vasquez-Yanes, C., and H. Smith. 1982. Phytochrome control of seed germination in the tropical rain forests' pioneer trees *Cecropia obtusifolia* and *Piper auritum*, and its ecological significance. *New Phytologist* **92**: 477–485. (Chap. 10)

Vitousek, P. M., P. R. Ehrlich, A. E. Ehrlich, and P. A. Matson. 1986. Human appropriation of the products of photosynthesis. *Bioscience* **36**: 368–373. (Preface)

Watterson, G. A. 1974. Models for the logarithmic species abundance distributions. *Theoretical Population Biology* **6**: 217–250. (Chaps. 5, 6)

Webb, T. III. 1988. Eastern North America. Pp. 385–414 in B. Huntley and T. Webb III, eds., *Vegetational History*. Kluwer, Dordrecht, The Netherlands. (Chap. 1)

Weiher, E., and P. Keddy. 1999. *Ecological Assembly Rules: Perspectives, Advances, Retreats*. Cambridge University Press, Cambridge, UK. (Chap. 1)

Weiner, J. 1990. Asymmetric competition in plant populations. *Trends in Ecology and Evolution* **5**: 360–364. (Chap. 3)

Weins, J. A. 1984. On understanding a non-equilibrium world: Myth and reality in community patterns and processes. Pp. 439–457 in D. R. Strong, D. Simberloff, L. G. Abele, and A. B. Thistle, eds., *Ecological Communities: Conceptual Issues and the Evidence*. Princeton University Press, Princeton, NJ. (Chap. 1)

Wheeler, Q. D., and J. Cracraft. 1997. Taxonomic preparedness: Are we ready to meet the biodiversity challenge? Pp. 435–446 in M. L. Reaka, D. E. Wilson, and E. O. Wilson, eds., *Biodiversity II*. Joseph Henry Press, Washington, DC. (Chap. 9)

White, M.J.D. 1978. *Modes of Speciation*. W. H. Freeman, San Francisco. (Chaps. 5, 9)

Whittaker, R. H. 1951. A criticism of the plant association and climatic climax concepts. *Northwest Science* **25**: 17–31. (Chap. 1)

Whittaker, R. H. 1956. Vegetation of the Great Smoky Mountains. *Ecological Monographs* **26**: 1–80. (Chap. 1)

Whittaker, R. H. 1965. Dominance and diversity in land plant communities. *Science* **147**: 250–260. (Chap. 1, 2, 5, 10)

Whittaker, R. H. 1967. Gradient analysis of vegetation. *Biological Reviews* **42**: 207–264. (Chap. 1)

Whittaker, R. H. 1975. *Communities and Ecosystems.* 2nd ed. Macmillan, New York. (Chap. 2)

Whittaker, R. H., and S. A. Levin, eds. 1975. *Niche Theory and Application.* Benchmark Papers in Ecology, vol. 3. Dowden, Hutchinson & Ross, Stroudsburg, PA. (Chap. 1)

Whitworth, W. A. 1934. *Choice and Chance.* Steichert, New York. (Chap. 2)

Williams, C. B. 1939. An analysis of four years' captures of insects in a light trap. Part I. General survey: Sex proportion; phenology; and time of flight. *Transactions of the Royal Entomological Society, London* **89**: 79–131. (Chap. 2)

Williams, C. B. 1940. An analysis of four years' captures of insects in a light trap. Part II. The effect of weather conditions on insect activity; and the estimation and forecasting of change in the insect population. *Transactions of the Royal Entomological Society, London* **90**: 227–306. (Chap. 2)

Williams, C. B. 1953. The relative abundance of different species in a wild animal population. *Journal of Animal Ecology* **22**: 14–31. (Chap. 2)

Williams, C. B. 1964. *Patterns in the Balance of Nature and Related Problems of Quantitative Biology.* Academic Press, London. (Chaps. 2, 6, 9)

Williamson, M. 1988. Relationship of species number to area, distance and other variables. Pp. 91–115 in A. A. Myers and P. S. Giller, eds., *Analytical Biogeography.* Chapman and Hall, New York. (Chap. 6)

Willis, J. C. 1922. *Age and Area: A Study of Geographical Distribution and Origin in Species.* Cambridge University Press, Cambridge, UK. (Chap. 1)

Wilson, E. O. 1995. *The Diversity of Life.* Harvard, University Press, Cambridge, MA. (Chap. 9)

Wilson, E. O., and F. M. Peter, eds. 1988. *Biodiversity.* National Academy Press, Washington, DC. (Preface)

Winemiller, K. O. 1996. Dynamic diversity in fish assemblages of tropical rivers. Pp. 99–134 in M. L. Cody and J. A. Smallwood, eds. *Long-Term Studies of Vertebrate Communities.* Academic Press, San Diego, CA. (Chap. 9)

Wright, S., and S. P. Hubbell. 1983. Stochastic extinction and reserve size: A focal species approach. *Oikos* **41**: 466–476. (Chap. 7)

Yodzis, P. 1986. Competition, mortality, and community structure. Pp. 480–491 in J. Diamond and T. J. Case, eds., *Community Ecology*. Harper and Row, New York. (Chap. 3)

Zhang, D.-Y., and K. Lin. 1997. The effects of competitive asymmetry on the rate of competitive displacement: How robust is Hubbell's community drift model? *Journal of Theoretical Biology* **188**: 361–367. (Chap. 10)

Index

all taxa biological inventory, 282

Belize forests, fit to theory, 142
biodiversity
  definition of, 3
  equilibrium in, 10
  fractal nature of, 259, 263
  Phanerozoic marine diversity, 274
biotic saturation
  of landscapes, 52
  of the BCI forest, 53
  in relation to zero-sum dynamics,
    54
body size, distribution of, 313
broken-stick hypothesis, 30, 39

character displacement, 15
Chesson and Warner's model, 65
  frequency dependence in, 65, 67
  relative species abundance
    effects in, 68
coexistence
  classical theory's preoccupation
    with, 11, 320
  and ecological drift, 10
  and environmental variance, 66
  and niche differentiation, 10
  and stochastic logistic growth, 69
commonness and rarity, 30, 132,
  138
competition-colonization trade-off,
  226
conservation biology, 202
  dispersal and conservation, 207
  incidence function, 203
  persistence function, 205
  reserve size/number (SLOSS)
    debate, 226
coordinated stasis hypothesis, 24
correlation length in biogeography,
  162, 175, 190

covariance of abundance on two
  islands, 209

demographic stochasticity, 6, 66, 69
density dependence, 55
dispersal-assembly perspective
  definition of, 8
  difficulty of falsifying, 24
  reconciliation with niche
    assembly, 26
  short history of, 21
dispersal limitation
  and metacommunity biodiversity,
    215, 218
  and species-area curves, 171, 173,
    189
  and species coexistence, 342:
    Tilman's argument, 344; Hurtt
    and Pacala's argument, 344
dominance-diversity curve
  in closed-canopy forests, 116,
    142, 144, 147, 294, 303, 334
  definition of, 36
  diverse examples of, 4, 302

ecological communities
  definition of, 5
  equilibrium, 20
  trophic organization of, 5
ecological drift
  absorbing case, definition of, 78
  definition of, 6, 57
  ergodic case, definition of, 78
  graphical representation of, 75
  single-species dynamics under:
    see local community dynamics
ecological guild, 5
ecological dominance deviations,
  338
ecological equivalence
  under fitness invariance rules,
    322, 339

371